Extreme Nonlinear Optics

Advanced Texts in Physics

This program of advanced texts covers a broad spectrum of topics which are of current and emerging interest in physics. Each book provides a comprehensive and yet accessible introduction to a field at the forefront of modern research. As such, these texts are intended for senior undergraduate and graduate students at the MS and PhD level; however, research scientists seeking an introduction to particular areas of physics will also benefit from the titles in this collection.

Martin Wegener

Extreme Nonlinear Optics

An Introduction

With 86 Figures, Numerous Examples,
and 29 Problems with Complete Solutions

 Springer

Professor Dr. Martin Wegener

Institut für Angewandte Physik
Universität Karlsruhe (TH)
Wolfgang-Gaede-Str. 1
76131 Karlsruhe, Germany
E-mail: martin.wegener@physik.uni-karlsruhe.de

ISSN 1439-2674

ISBN 3-540-22291-x   Springer-Verlag Berlin Heidelberg New York

Library of Congress Control Number: 2004109596

Springer is a part of Springer Science+Business Media

springeronline.com

© Springer-Verlag Berlin Heidelberg 2005
Printed in Germany

Typesetting: by the author and F. Herweg EDV Beratung using a Springer TeX macro package
Cover design: *design & production* GmbH, Heidelberg

Printed on acid-free paper   SPIN 10984680   56/3141/jl   5 4 3 2 1 0

This book is dedicated to Karin, Pauline, and Henriette.

Preface

In 2001 I was invited to give a series of lectures on our work on strong-field excitation of semiconductors at a 2003 summer school in Erice (Sicily). I enthusiastically started preparing the lectures and soon developed the idea to put our own work into a much broader context. When reading the original publications from diverse fields such as solid-state physics, atomic physics, relativistic physics, particle physics and metrology, I was at first overwhelmed by the amount of material and its diversity. A whole new world of nonlinear optics opened up to me. Then, I realized many similarities in the physics, despite tremendous differences in the jargon of the various scientific communities. Many of these exciting and novel examples of *Extreme Nonlinear Optics* had not been described in any textbook – so I decided to write this book.

It gives an introduction into this vibrant field from the viewpoint of an experimental physicist who cares about theory. Chapters one to six deliver the basics. Whenever necessary, crucial results from traditional nonlinear optics are briefly recapitulated. More than twenty problems with detailed solutions at the end of the book, as well as many examples, are an integral part of this introduction. Chapters seven and eight review the field and guide the reader towards its current state-of-the-art.

Thus, this book should be helpful for students of physics or electrical engineering in courses/seminars on nonlinear optics, quantum optics or quantum electronics as well as for young researchers entering the field of extreme nonlinear optics. Experts in one of the subfields might be surprised by the interconnections and parallels between the disciplines.

Karlsruhe, June 2004

Martin Wegener

Contents

1

Introduction

With the invention of the laser [1, 2] in general, and with the realization of the ruby laser by Maiman in 1960 [3] in particular, the field of optics soon entered the new era of *nonlinear optics* in 1961 [4]. In this regime, the optical properties of materials are no longer independent of the intensity of light – as was believed for hundreds of years before – but rather change with the light intensity, giving rise to a wealth of new phenomena, effects and applications. Today, nonlinear optics has entered our everyday life in many ways and has also been the basis for numerous new developments in spectroscopy and laser technology. Indeed, from the moment of birth of nonlinear optics, laser physics and nonlinear optics have been intimately related to each other.

1.1 "Traditional" Nonlinear Optics – Extreme Nonlinear Optics

Within *"traditional" nonlinear optics*, the absolute changes of the optical properties are tiny if one follows them versus time on a timescale of a cycle of light. This simple fact is the basis of many concepts and approximations of "traditional" nonlinear optics – as described in a number of excellent textbooks [5–10]. Over the years, however, lasers have improved in many ways, especially in terms of the accessible peak intensities and in terms of the minimum pulse duration available. These days, about 40 years after the invention of the laser, the shortest optical pulses generated are about one and a half cycles of light in duration (see Fig. 1.1). This comes close to the ultimate limit of a single optical cycle. By virtue of mode-locking [11] and specifically of self-mode-locking [12] of solid-state lasers, such pulses can even be generated directly from the laser oscillator. Moreover, thanks to the concept of chirped-pulse amplification (CPA) [16, 17], amplified laser pulses with focused peak intensities in the range of 10^{22} W/cm^2 [18] are available in some laboratories (see Fig. 1.2). In ten years from now, this gigantic number could possibly be further increased by another several orders of magnitude. As a result of this, today's light intensities can lead to substantial or even to extreme changes on the timescale of light. We will see later in this book that this somewhat vague statement can be specified by saying:

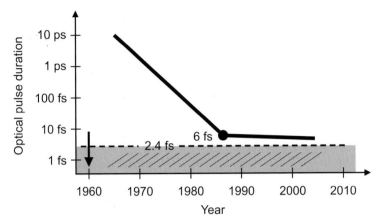

Fig. 1.1. Development of the minimum available laser pulse duration t_{FWHM} versus year (schematically). The ruby laser came into operation in 1960 (arrow) and only five years later, 10-ps pulses were available. Thereafter, the pulse duration decayed almost exponentially for two decades. Not too much happened in terms of duration after the 1987 world record of Shank et al. [13] based on a dye laser and a dye amplifier. However, the subsequent solid-state revolution [14, 15] led to an enormous progress in terms of reliability. The shortest possible *optical* pulse duration of 2.4 fs is indicated (this is equivalent to 1.3 cycles of light at 2.25 eV center photon energy, see Problem 2.2). In this figure, we do not consider sub-femtosecond extreme ultraviolet pulses generated via high-harmonic generation (see Example III).

> Whenever an energy associated with the light intensity becomes comparable to or even larger than a characteristic energy of the material or system under investigation, the laws of "traditional" nonlinear optics fail and something new is expected to happen.

We want to call this regime *extreme nonlinear optics* or carrier-wave nonlinear optics. The latter is more precise, the notion *extreme nonlinear optics* is popular as it sounds more "sexy". Depending on the problem and/or system under consideration, the energy associated to the light intensity, I, can be one of the five energies:

- Rabi energy $\hbar\Omega_{\mathrm{R}} \propto \sqrt{I}$
- Ponderomotive energy $\langle E_{\mathrm{kin}} \rangle \propto I$
- Bloch energy $\hbar\Omega_{\mathrm{B}} \propto \sqrt{I}$
- Cyclotron energy $\hbar\omega_{\mathrm{c}} \propto \sqrt{I}$
- Tunneling energy $\hbar\Omega_{\mathrm{tun}} \propto \sqrt{I}$.

The characteristic energy of the system under investigation can be one of the three energies:

- Carrier photon energy $\hbar\omega_0$ (or transition energy $\hbar\Omega$)
- Binding energy E_{b}
- Rest energy $m_0 c_0^2$.

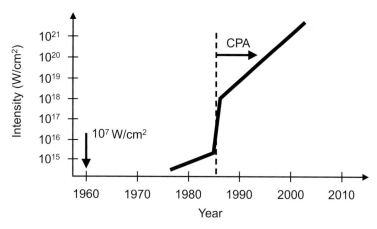

Fig. 1.2. Development of the maximum available focused laser intensity I versus year (schematically). A focused ruby laser in 1960 (see arrow) led to approximately ten million W/cm^2. The birth of chirped-pulse amplification (CPA) in 1985 is indicated. Thereafter, the intensity has increased approximately exponentially for nearly two decades. Further increase is expected in the future.

The following table gives an overview.

	$\hbar\Omega_R$	$\langle E_{kin}\rangle$	$\hbar\Omega_B$	$\hbar\omega_c$	$\hbar\Omega_{tun}$
$\hbar\omega_0$	3, 7.1, 7.2	4.2, 7.3	4.3	4.4, 4.5, 8.2	5.2, 5.3, 5.4, 8.1
E_b		5.2, 5.3, 5.4, 8.1			
$m_0 c_0^2$		4.4, 4.5, 8.2		4.5	

For example, the case $\langle E_{kin}\rangle \gtrsim \hbar\omega_0$ will be discussed in Sects. 4.2 and 7.3. Missing entries correspond to irrelevant cases.

This book would not have been written if the resulting modifications of "traditional" nonlinear optics had turned out to be minor corrections or details. Quite in contrast to this, qualitatively new effects arise in extreme nonlinear optics. Let us briefly address some of them here to give a first flavor of what this book is about.

Example I

In most of today's green pump "lasers", the green light is not actually generated directly by a laser. Instead, a near-infrared laser is doubled in its light frequency using a so-called second-harmonic generation (SHG) crystal. Within "traditional" nonlinear optics, such frequency doubling is only possible if the medium lacks inversion symmetry. It would, e.g., not work in glass. In extreme nonlinear optics, we carefully have to distinguish between the notions of *frequency doubling* on the one hand and *second-harmonic generation* on the other hand. Whenever one of them interferes with another contribution, the distinction between the two becomes actually relevant. Both can occur in media *with inversion symmetry.*

Frequency doubling refers to a peak in the spectrum at twice the laser carrier frequency. It can, e.g., occur in semiconductors excited by intense few-cycle light pulses via *third-harmonic generation in the disguise of second-harmonic generation.* Experimentally, this mechanism can be as intense as "traditional" frequency doubling via SHG. Another corresponding mechanism is *carrier-wave Rabi flopping.* Here, a sideband of the laser carrier frequency can spectrally coincide with the second harmonic.

Second-harmonic generation always leads to frequency doubling but it is more stringent than frequency doubling. It requires that the carrier frequency of the radiation is given by twice the laser carrier frequency. *Conical second-harmonic generation* can lead to true SHG with a particular conical emission profile in isolators with inversion symmetry or in gases. Another mechanism leading to true second-harmonic generation in the presence of inversion symmetry is addressed in Example IV.

Example II

Most laser pulses are well described by a (slowly varying) temporal envelope and a carrier-wave oscillation. The phase in between the two, the so-called *carrier-envelope offset phase*, is practically irrelevant in "traditional" nonlinear optics. Why is this? As we have argued above, the nonlinear optical changes are tiny on a timescale of a cycle of light in "traditional" nonlinear optics and, hence, one can forget about the carrier-wave oscillation and describe everything exclusively in terms of the envelope. Clearly, at this point the carrier-envelope offset (CEO) phase drops out of the picture. As the instantaneous light intensity is proportional to the square modulus of the electric-field envelope, one can express most aspects of "traditional" nonlinear optics in terms of the light intensity. This procedure is not valid in the regime of extreme nonlinear optics and, thus, in this sense, it is meaningful to say that *the electric field governs the behavior rather than the light intensity.* Hence, the CEO phase has a substantial influence on the outcome of an experiment. It will, e.g., allow us to distinguish between mere frequency doubling and true second-harmonic generation (see Example I). One could, alternatively, also turn the story around and define extreme nonlinear optics by a dependence of the nonlinear optical signals on the CEO phase.

We will see that the pulse-to-pulse change of the CEO phase, which is proportional to the CEO frequency, has turned out to be a key in building *optical clocks.* Such optical clocks can be so precise (relative precision around 10^{-15}) that one might be able to observe the change of fundamental "constants" – a discussion that started with a 1937 paper of Dirac [19]. One could, e.g., possibly measure the variation of the fine structure "constant" in the laboratory (via spectroscopy of hydrogen) on a laboratory timescale rather than on a cosmological scale [20, 21].

Example III

As already pointed out above, optical pulses with a duration of less than an optical cycle are generally not possible. For example, for red light with about $2\,e$V photon energy, a cycle of light corresponds to a period of about two femtoseconds. If one wants to generate yet shorter pulses, the carrier frequency must be larger, bringing us into the ultraviolet (UV) or extreme UV region. Unfortunately, laser oscillators in this regime (apart from free electron lasers) are not readily available today. One of the success stories of extreme nonlinear optics is the *generation of high harmonics in gases*. At this point, "high" refers to the 101st or to the 247th harmonic of a laser. Under appropriate conditions, such high harmonics can lead to single attosecond pulses or pulse trains. Amazingly, corresponding experimental setups can even fit into a regular-size laboratory, in turn allowing such pulses to be applied in spectroscopy or, possibly, in extreme UV lithography of future computer chips.

Example IV

The interaction of a light wave with an electron in vacuum does not lead to any nonlinear optical response within "traditional" nonlinear optics. In extreme nonlinear optics, the cyclotron energy associated with the light field can become comparable to the carrier photon energy, in which case the magnetic part of the Lorenz force becomes important and *relativistic effects* occur. For example, true second-harmonic generation (see Example I) can arise – even though an electron in vacuum clearly has inversion symmetry.

Example V

Traditional nonlinear optics invalidates the superposition principle of light waves in nonlinear optical media. Yet, in vacuum, the superposition principle of the Maxwell equations still holds. This statement is no longer correct in extreme nonlinear optics. Here, *photogeneration of electron–positron pairs* can lead to an effective photon–photon interaction and, hence, to third-harmonic generation or even to Rabi oscillations, i.e., to a momentary inversion of the Dirac sea.

Range of intensities

To get a feeling for the involved laser intensities, let us consider the enormous span of orders of magnitude shown in Table 1.1 ranging from the very dim to the extremely bright. Here, the light intensity I covers more than *fifty orders of magnitude* from $I = 10^{-23}$ W/cm^2 to $I = 10^{+30}$ W/cm^2. The visible part of the black-body radiation at room temperature sets a lower bound as to how dark it can get in a room (see Problem 1.1). The sun's light intensity on the earth is 22 orders of magnitude larger than that. Ten Watts of power from a continuous-wave laser with a one-millimeter beam diameter, corresponding to a light intensity of 10^3 W/cm^2, hurt if you stick your finger in the beam. Try it! Another nine orders of magnitude above, extreme nonlinear optics in solids starts. Yet another two orders of magnitude further and extreme nonlinear optics in atoms takes place. Another three orders of magnitude bring us to relativistic effects of electrons in vacuum. Five more orders of magnitude might lead to the observation of nonlinear optics in vacuum, yet five orders of magnitude more to the generation of Unruh radiation, which is somewhat

Table 1.1. Light intensities I (in units of W/cm^2) from the very dim to the extremely bright.

10^{+30}	$\rightarrow$ generation of real electron–positron pairs from vacuum
10^{+28}	$\rightarrow$ electron acceleration by light comparable to edge of black hole
10^{+26}	
10^{+24}	$\rightarrow$ nonlinear optics of the vacuum ?
10^{+22}	
10^{+20}	$\rightarrow$ photonuclear fission – light splits nuclei
10^{+18}	$\rightarrow$ relativistic nonlinear optics of vacuum electrons
10^{+16}	
10^{+14}	$\rightarrow$ electrostatic tunneling of electrons from atoms
10^{+12}	$\rightarrow$ Rabi flopping in semiconductors becomes optical
10^{+10}	
10^{+8}	
10^{+6}	$\rightarrow$ laser intensity in the first experiment on nonlinear optics in 1961
10^{+4}	
10^{+2}	$\rightarrow$ a continuous-wave laser of that intensity hurts
1	$\rightarrow$ total intensity of the sun on the earth's surface (10^{-1} W/cm^2)
10^{-2}	$\rightarrow$ thermal radiation from a human
10^{-4}	
10^{-6}	
10^{-8}	
10^{-10}	$\rightarrow$ total intensity of the cosmic 2.8 K background radiation
10^{-12}	
10^{-14}	
10^{-16}	
10^{-18}	
10^{-20}	
10^{-22}	$\rightarrow$ visible intensity in a "dark" room at 300 K (10^{-23} W/cm^2)

similar to Hawking radiation from the edge of a black hole. At the gargantuan intensity of $I = 10^{30}$ W/cm^2, the Schwinger intensity, the potential drop of the laser electric field over the electron Compton wavelength is comparable to the electron rest energy.

Problem 1.1. Calculate the light intensity in a "dark" room held at room temperature ($T = 300$ K).

1.2 How to Read this Book?

Within this book, we will describe these and other phenomena starting from a rather elementary level. Even readers not familiar with "traditional" nonlinear optics at all should be able to follow the lines of argument as crucial results of "traditional" nonlinear optics are repeated.

You can read this book from the beginning to the end. But there are alternative ways to use it. For example, some readers might be familiar with the Lorentz oscillator model and the Drude model, which are frequently used to describe the linear optical properties of materials on an elementary level. The two self-contained chapters on "The Lorentz Oscillator Model and Beyond ..." describing *bound–bound transitions* and on "The Drude Free-Electron Model and Beyond ..." on *unbound–unbound transitions* start from these simple models and gradually and pedagogically guide the reader into extreme nonlinear optics using pretty "harmless" mathematics. The following chapter "Lorentz Becomes Drude: ..." discusses the mixed case of *bound–unbound transitions*. A number of examples and problems help the reader to become familiar with the concepts and the relevant quantities and numbers. Depending on your background, an introduction into few-cycle laser pulses and their properties can be found in the chapter on "Selected Aspects of Few-Cycle Laser Pulses and Nonlinear Optics". More advanced readers might already be familiar with these basics and decide to jump to later chapters. There, we have attempted to give a rather concise overview about experiments on extreme nonlinear optics in various systems, ranging from electrons bound in solids to electrons bound in atoms to electrons in vacuum.

It is strongly recommended to consider the problems while progressing with the text in order to test whether you are actually familiar with the material. These problems are an integral part of this book. Some of them are easy, others are rather difficult, some ask for a qualitative answer, others require a few pages of mathematics. Required fundamental constants can be found in the list of symbols. The detailed solutions of all problems are given at the end of this book.

Whenever you become lost with any of the mathematical symbols, the list of symbols at the end of the book comes to the rescue. We have avoided the use of the same symbol for different quantities as much as possible. Note that, as a result, some symbols are not identical to what is most commonly used in a specific community.

2

Selected Aspects of Few-Cycle Laser Pulses and Nonlinear Optics

The basic principle of a laser is simple: It consists of a resonator (see Fig. 2.1) and a light amplifier – the gain medium. For the purpose of this book, the quantum optical aspects of the light field are not important, hence, it is sufficient to consider the well-known Maxwell equations of electrodynamics [22].

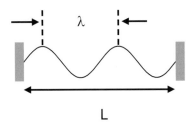

Fig. 2.1. Scheme of a laser resonator, consisting of two mirrors separated by length L. A snapshot of a single mode ($N = 4$) of the electric field at wavelength λ is shown.

2.1 Maxwell Equations

In S.I. units the Maxwell equations are given by

$$\nabla \cdot \boldsymbol{D} = \rho \tag{2.1}$$

$$\nabla \times \boldsymbol{E} = -\frac{\partial \boldsymbol{B}}{\partial t} \tag{2.2}$$

$$\nabla \cdot \boldsymbol{B} = 0 \tag{2.3}$$

$$\nabla \times \boldsymbol{H} = +\frac{\partial \boldsymbol{D}}{\partial t} + \boldsymbol{j} \,. \tag{2.4}$$

ρ is the electric charge density and j the electric current density. In media, the relation between the E-field and the D-field is given by[1]

$$E = \frac{1}{\varepsilon_0} (D - P) \,, \tag{2.5}$$

with the macroscopic polarization P, and, similarly, for the B-field and the H-field the relation

$$B = \mu_0 (H + M) \,, \tag{2.6}$$

with the magnetization M. For the materials relevant in the context of this book, $M = 0$ holds, and (2.6) simplifies to

$$B = \mu_0 H \,. \tag{2.7}$$

In linear optics, one has

$$P = \varepsilon_0 \chi E \,, \tag{2.8}$$

with the linear optical susceptibility χ. In this case, (2.5) simplifies to

$$D = \varepsilon_0 \varepsilon E \,, \tag{2.9}$$

with the relative dielectric function $\varepsilon = 1 + \chi$. The Maxwell equations can be rewritten into the known wave equation[2] for the E-field

$$\Delta E - \frac{1}{c_0^2} \frac{\partial^2 E}{\partial t^2} = +\mu_0 \frac{\partial^2 P}{\partial t^2} \,, \tag{2.10}$$

or, using (2.8), into

$$\Delta E - \frac{1}{c^2} \frac{\partial^2 E}{\partial t^2} = 0 \,, \tag{2.11}$$

with the medium velocity of light $c = c_0/n$, which is slower than the vacuum velocity of light $c_0 = 1/\sqrt{\epsilon_0 \mu_0} = 2.998 \times 10^8$ m/s by a factor identical to the (generally complex) refractive index n with

$$n = \sqrt{\varepsilon} \,. \tag{2.12}$$

2.2 The Light Intensity

Our eyes and most detectors are not sensitive to the electric field itself but to the number of photons that hit the detector per unit time. In other words, classically speaking: They are sensitive to the cycle-average of the modulus of the Poynting

---

[1] $\epsilon_0 = 8.8542 \times 10^{-12}$ A s V^{-1}m^{-1} and $\mu_0 = 4\pi \times 10^{-7}$ V s A^{-1}m^{-1}.

[2] Coming from Karlsruhe, we just have to remind you that it was Karlsruhe where H. Hertz found the first experimental evidence for electromagnetic waves in the year 1887.

vector $S = E \times H$. For plane waves in vacuum one has $|B| = |E|/c_0$ or equivalently $|E| = |H|\sqrt{\frac{\mu_0}{\varepsilon_0}}$, with the vacuum impedance

$$\sqrt{\frac{\mu_0}{\varepsilon_0}} = 376.7301\,\Omega\,, \tag{2.13}$$

leading to

$$S = |S| = \sqrt{\frac{\varepsilon_0}{\mu_0}}\,|E|^2\,, \tag{2.14}$$

which generally varies with time. For an electric field according to, e.g.,

$$|E(t)|^2 = \tilde{E}_0^2 \cos^2(\omega_0 t + \phi)\,, \tag{2.15}$$

the *light intensity* I, which is defined as the cycle-average[3] of the modulus of the *Poynting vector*, becomes[4]

$$I = \langle S \rangle = \frac{1}{2}\sqrt{\frac{\varepsilon_0}{\mu_0}}\,\tilde{E}_0^2\,. \tag{2.16}$$

Note that the intensity I does not depend on ϕ.

▶ **Example 2.1.** An electric field of $\tilde{E}_0 = 4 \times 10^9$ V/m in vacuum corresponds to an intensity of $I = 2.1 \times 10^{12}$ W/cm^2. For comparison: This intensity corresponds to concentrating the power of a thousand nuclear power plants with a power of 2 GW each onto an area comparable to your finger tip – for a very short time. For the same electric field, the peak of the **B**-field envelope is $\tilde{B}_0 = \mu_0\sqrt{\varepsilon_0/\mu_0}\,\tilde{E}_0 = \tilde{E}_0/c_0 = 13.3$ T. ◀

Problem 2.1. A light field with $\tilde{E}_0 = 4 \times 10^9$ V/m propagates from air into a dielectric with $\epsilon = 10.9$. Suppose that reflections are completely suppressed via an ideal antireflection (AR) coating. What is $\tilde{E}_0$ inside the dielectric?

2.3 Electric Field in a Laser Resonator

Solutions of the wave equation (2.11) are, e.g., plane waves with

$$E(r,t) = E_0 \cos(Kr - \omega t - \phi) = \frac{E_0}{2}\exp(i(Kr - \omega t - \phi)) + \text{c.c.}\,, \tag{2.17}$$

which have to obey the *dispersion relation of light*

---

[3] Remember that $\langle\cos^2(\omega_0 t + \phi)\rangle = 1/2$.
[4] Within a dielectric, ϵ_0 has to be replaced by $\epsilon_0\,\epsilon$ in this relation.

$$\frac{\omega}{|K|} = c = \frac{c_0}{n(\omega)} \qquad (2.18)$$

for the frequency ω and the wavevector of light K.

For the resonator shown in Fig. 2.1, we have the superposition of left-going and right-going waves, i.e., a standing wave, such that the electric field has nodes at the two mirrors, $E(z = 0, t) = E(z = L, t) = 0$. Thus, the *length L of the resonator* has to be an integer multiple, let us say N, of half the wavelength of light λ:

$$L = N \frac{\lambda}{2}. \qquad (2.19)$$

With the dispersion relation of light (2.18) and with $|K| = 2\pi/\lambda$, this can be rewritten into

$$\omega_N = N \Delta\omega, \qquad (2.20)$$

with the mode spacing

$$\Delta\omega = c \frac{\pi}{L}. \qquad (2.21)$$

▶ **Example 2.2.** For a resonator with length $L = 1.5\,\mathrm{m}$ and with $c = c_0$ we obtain $\Delta\omega = 2\pi \times 100\,\mathrm{MHz}$, which is within the radio-frequency (RF) regime.   ◀

The superposition principle tells us that any linear combination of these eigensolutions is also a solution of the resonator problem and we can write the general solution of standing waves in the resonator as

$$E(r, t) = \sum_{N=1}^{\infty} 2 E_0^N \sin(K_{Nz}z) \sin(\omega_N t + \varphi_N). \qquad (2.22)$$

Details depend on the amplitudes E_0^N, the phases φ_N of the modes with $N = 1-\infty$ and also on the dispersion relation (2.18) that connects the K_N and the ω_N. Let us consider three cases, (a)–(c) in Fig. 2.2, in which we study one component of the electric-field vector, $E(t) = E_{x,y}(r = \mathrm{const.}, t)$, at a fixed point in space within the cavity. For the sake of simplicity we assume that the mode amplitudes are either constant or zero (which mimics the finite bandwidth of the gain medium), i.e., $E_0^N = E_0$ for all frequencies in the interval $[\omega_0 - \delta\omega/2, \omega_0 + \delta\omega/2]$ and $E_0^N = 0$ else. We choose $\delta\omega/\omega_0 = 0.6$.

In (a) we consider many modes N with random phases φ_N, $c = \mathrm{const}$. This leads to an electric field that looks like noise with some average intensity (Fig. 2.2(a)). This situation corresponds to a multimode continuous-wave (cw) laser – a bad cw laser. We conclude that a good cw laser must only work on a single mode.

In (b) all the phases φ_N are equal – they are locked – in which case we can set them to zero, $c = \mathrm{const}$. A periodic train of *identical* pulses results (Fig. 2.2(b)). The duration of the individual pulses is inversely proportional to the width of the frequency

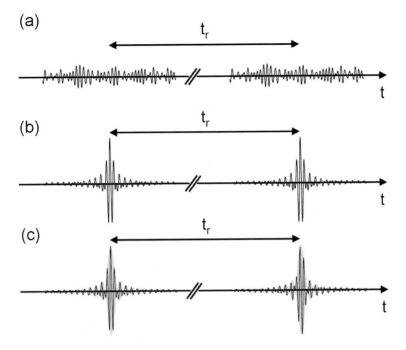

Fig. 2.2. Electric field versus time t in the middle of the laser cavity according to (2.22). **(a)** Random phases φ_N, $c = $ const., **(b)** $\varphi_N = 0$ for all N, $c = $ const., **(c)** $\varphi_N = 0$ for all N, $c = c(\omega_N) \neq$ const. Note that (b) and (c) have been demagnified with respect to (a) in the vertical direction by a factor of about 10^6.

interval $\delta\omega$. How can one realize this locking of the modes experimentally? By active or passive modulation of the resonator properties with frequency $\Delta\omega$, which is called *mode-locking*! Such modulation of the mode with frequency ω_N leads to sidebands at $\omega_N + \Delta\omega = \omega_{N+1}$ and $\omega_N - \Delta\omega = \omega_{N-1}$ for all N. This couples all the modes, hence it locks their phases φ_N, and it leads to a perfectly equidistant spacing of the modes in frequency space.

In (c) we give up the unrealistic assumption of a constant velocity of light c in the resonator, but the modes will still be equidistant in frequency. The corresponding time-domain behavior of (2.22) is schematically shown in Figs. 2.2(c) and 2.3. The pulses are *not identical* under these conditions. In one roundtrip, a shift of the phase between the envelope and the carrier wave results from the fact that the *group velocity* v_{group} at frequency ω_0 (the velocity of the envelope)

$$v_{\text{group}} = \frac{d\omega}{dK}, \qquad (2.23)$$

with $K = |\mathbf{K}|$ and the *phase velocity* v_{phase} (the velocity of the carrier wave at frequency ω_0)

$$v_{\text{phase}} = c = \frac{\omega}{K} \qquad (2.24)$$

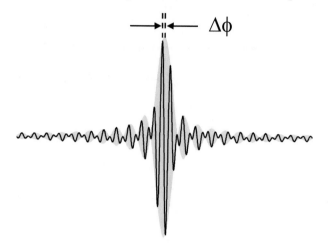

Fig. 2.3. Magnification of the electric field versus time from the RHS of Fig. 2.2(c). The gray area corresponds to the electric-field envelope. In this example, the CEO phase is $\phi = +\pi/2$, which corresponds to a pulse-to-pulse phase shift of $\Delta\phi = +\pi/2$ in Fig. 2.2(c).

are no longer identical. We can define a corresponding *carrier-envelope offset fre-quency* f_ϕ that is generally different from the repetition frequency $f_r = 1/t_r = \Delta\omega/(2\pi)$. From Fig. 2.2(c) it becomes clear that the electric field according to (2.22) can alternatively be expressed[5] as

$$E(t) = \sum_{N=-\infty}^{+\infty} \tilde{E}(t - N\,t_r)\cos\left(\omega_0(t - N\,t_r) + N\,\Delta\phi + \phi\right). \qquad (2.25)$$

The cosine-term is the *carrier-wave oscillation* with carrier frequency ω_0, the pref-actor $\tilde{E}$ is called the *envelope* of the pulse (gray areas in Fig. 2.2). t_r is the roundtrip time, $\Delta\phi$ the pulse-to-pulse phase slip, and ϕ an overall phase. $(N\,\Delta\phi + \phi)$ is under-stood as mod 2π, i.e. for all integers N, the term is an element of the interval $[0, 2\pi]$. Later, we will only consider *one pulse out of the pulse train* according to (2.25), e.g., the one with $N = 0$, which leads to an electric field of

$$E(t) = \tilde{E}(t)\cos(\omega_0 t + \phi), \qquad (2.26)$$

with the so-called *carrier-envelope offset (CEO) phase*[6] ϕ. The CEO phase of a single pulse has to be distinguished from the well-known relative optical phase between two beams or pulses, e.g., in a Michelson interferometer.

---

[5] Note that the choice of the carrier frequency ω_0 is somewhat arbitrary, especially if the pulses are chirped. Often, one chooses ω_0 as the center of mass of the frequency spectrum of the laser pulses.

[6] The CEO phase is sometimes also called the absolute optical phase.

What is the frequency-domain analogue of this behavior? We compute the Fourier transform of the electric field, $E(\omega)$, via

$$E(\omega) = \frac{1}{\sqrt{2\pi}} \int_{-\infty}^{+\infty} E(t)\, e^{+i\omega t}\, dt \tag{2.27}$$

$$= \frac{1}{\sqrt{2\pi}} \int_{-\infty}^{+\infty} \sum_{N=-\infty}^{+\infty} \tilde{E}(t - N\, t_r) \cos\Big(\omega_0(t - N\, t_r) + N\,\Delta\phi + \phi\Big) e^{+i\omega t}\, dt\,.$$

The cosine-term can be written according to

$$\cos(\omega_0 t + \ldots) = \frac{1}{2}\left(e^{+i(\omega_0 t + \ldots)} + e^{-i(\omega_0 t + \ldots)}\right)\,. \tag{2.28}$$

The exponential with the "minus" sign leads to a peak in $E(\omega)$ at positive frequencies ω, the term with the "plus" sign to a peak at negative ω. The latter is omitted at this point (the range of validity is discussed in Example 2.4) and we get

$$E(\omega) = \frac{1}{\sqrt{2\pi}} \int_{-\infty}^{+\infty} \sum_{N=-\infty}^{+\infty} \tilde{E}(t - Nt_r) \frac{1}{2} e^{-i(\omega_0(t - N\,t_r) + N\,\Delta\phi + \phi)}\, e^{+i\omega t}\, dt$$

$$= \frac{1}{2}\left(\sum_{N=-\infty}^{+\infty} e^{-i(N(\Delta\phi - \omega_0 t_r) + \phi)} \left(\frac{1}{\sqrt{2\pi}} \int_{-\infty}^{+\infty} \tilde{E}(t - Nt_r)\, e^{+i(\omega - \omega_0)t}\, dt\right)\right)$$

$$= \frac{1}{2}\left(\sum_{N=-\infty}^{+\infty} e^{-i(N(\Delta\phi - \omega_0 t_r) + \phi)}\, e^{+iN(\omega - \omega_0)t_r}\right)$$

$$\times \underbrace{\left(\frac{1}{\sqrt{2\pi}} \int_{-\infty}^{+\infty} \tilde{E}(t')\, e^{+i(\omega - \omega_0)t'}\, dt'\right)}_{=:\ \tilde{E}_{\omega_0}(\omega - \omega_0)}$$

$$= \frac{e^{-i\phi}}{2}\left(\sum_{N=-\infty}^{+\infty} e^{-iN(\Delta\phi - \omega t_r)}\right) \tilde{E}_{\omega_0}(\omega - \omega_0)\,. \tag{2.29}$$

From the second to the third line we have substituted $t' = (t - Nt_r)$. $\tilde{E}_{\omega_0}(\omega - \omega_0)$ is the envelope of the optical spectrum[7]. With an optical spectrometer[8] one usually

---

[7] In order to distinguish this envelope at carrier frequency ω_0 from other envelopes that will occur in this book, we have introduced the index ω_0.

[8] The full width at half-maximum (FWHM), $\delta\omega$, of the intensity spectrum multiplied by the FWHM of the temporal intensity profile, δt, is the duration–bandwidth product $\delta\omega\,\delta t$. One obtains $\delta\omega\,\delta t \geq 2\pi \times 0.4413$ for a Gaussian, i.e., for a $\exp(-t^2)$ pulse, $\delta\omega\,\delta t \geq 2\pi \times 0.8859$ for a $\mathrm{sinc}^2(t) = (\sin(t)/t)^2$ pulse, $\delta\omega\,\delta t \geq 2\pi \times 0.3148$ for a $\mathrm{sech}^2(t) = 1/\cosh^2(t)$ pulse, and $\delta\omega\,\delta t \geq 2\pi \times 0.1103$ for a one-sided exponential, i.e., for a $\Theta(t)\exp(-t)$ pulse [23]. The latter is the absolute minimum of the product $\delta\omega\,\delta t$ for any pulse shape. For all these cases, the equality applies for zero chirp.

measures the intensity spectrum $\propto |\tilde{E}_{\omega_0}(\omega - \omega_0)|^2$ (which neither depends on ϕ nor on $\Delta\phi$). The spectrum is modulated by the sum over the exponentials in the last line of (2.29). The significant values of ω in this sum are those for which the terms for, e.g., N and $(N + 1)$ add constructively, i.e., for which we have $(\omega t_r - \Delta\phi) = M\,2\pi$ with integer M. This yields an *equidistant ladder of angular frequencies* ω_M with

$$\omega_M = M\,\frac{2\pi}{t_r} + \frac{\Delta\phi}{t_r}\,. \tag{2.30}$$

Finally, we can convert from angular frequencies to frequencies via $\omega_M = 2\pi\,f_M$ and obtain

$$f_M = M\,f_r + f_\phi\,, \tag{2.31}$$

with the *repetition frequency* $f_r = 1/t_r = \Delta\omega/(2\pi)$ (see (2.21))

$$f_r = \frac{c}{2L} \tag{2.32}$$

and the *carrier-envelope offset frequency*

$$f_\phi = f_r\,\frac{\Delta\phi}{2\pi} \le f_r\,. \tag{2.33}$$

If $f_\phi \ne 0$, this frequency comb has a certain offset frequency [24–28]. If one can arrange for $f_\phi = 0$, on the other hand, the eigenfrequencies form a ladder of equidistant frequencies starting at zero frequency with $M = 0$. These findings are summarized in Fig. 2.4.

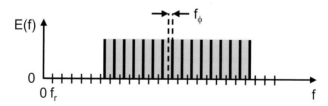

Fig. 2.4. Scheme of the frequency-domain analogue, $E(f)$, of the temporal behavior, $E(t)$, shown in Fig. 2.2(c). The spectrum exhibits peaks at the frequencies $f_M = M\,f_r + f_\phi$ with integer M, i.e., the equidistant frequency comb is upshifted by the carrier-envelope offset frequency f_ϕ. The actual $E(f)$ corresponding to Fig. 2.2(c) contains more than 10^6 densely spaced peaks within the optical spectrum (indicated by the gray area).

▶ **Example 2.3.** Let us compute the carrier-envelope offset frequency f_ϕ for the following parameters: $L = 1.5\,\text{m}$, $v_{\text{group}} = 99.9999\%\,c_0$, $v_{\text{phase}} = c_0$, $\hbar\omega_0 = 1.5\,\text{eV}$ $\Leftrightarrow 2\pi/\omega_0 = 2.8\,\text{fs}$. The difference, Δt, between the round-trip group delay time and the phase delay time is $\Delta t = 2L/v_{\text{group}} - 2L/v_{\text{phase}}$. With $1/(1 - x) \approx (1 + x)$ for

$x \ll 1$ we get $\Delta t \approx 2L/c_0 \times ((1 + 10^{-6}) - 1) = 10\,\text{ns} \times 10^{-6} = 10\,\text{fs}$. This leads to $\Delta\phi = (10\,\text{fs}/(2.8\,\text{fs}) \times 2\pi) \bmod 2\pi = (3.57 \times 2\pi) \bmod 2\pi = 0.57 \times 2\pi$. With (2.33) and with $f_r = 100\,\text{MHz}$ from Example 2.2 $\Rightarrow f_\phi = 57\,\text{MHz}$.   ◄

The time standard for one second within the S.I. system is related to a frequency of $9\,192\,631\,770\,\text{Hz} \approx 9\,\text{GHz}$. This frequency can rather easily be locked to the repetition frequency f_r of the oscillator, which is typically around $100\,\text{MHz}$ (see Example 2.2). If the pulse-to-pulse phase shift $\Delta\phi$ can be stabilized to $\Delta\phi = 0$ (Sect. 2.6), the frequency comb starts at zero frequency and looks *much like a ruler for frequencies* where one simply counts the number of millimeter markers to measure a length. This allows us to connect the time standard for one second in the GHz regime to optical frequencies at hundreds of THz. Previously, this required a very (!) complicated procedure (see references given in Refs. [26, 28]). The corresponding important implications for metrology (i.e., for using femtosecond lasers as frequency standards) are nicely discussed in rather recent review articles [26–28]. Some authors even speculate that this increased precision in measuring time might lead to experiments in which one could possibly observe the temporal variation of fundamental "constants" versus time on laboratory timescales [29–31]. If one could, e.g., measure the atomic Rydberg constant with increased frequency precision (relative precision of 10^{-15} or better), this value could be related to the fine-structure constant. Performing such an experiment today and comparing it to the result one year or several years later might reveal a difference. Strange but true: The ultrafast becomes useful for the ultraslow or ultraprecise, respectively. Relative precisions down to 10^{-18} are anticipated [32].

Let us finally note that one must be cautious with the choice of the electric-field envelope $\tilde{E}(t)$ according to (2.26). In actual experiments, the optical spectrum does not contain zero-frequency (dc) components. According to the Maxwell equations, zero-frequency components are not radiated at all, low-frequency components are not efficiently radiated. Furthermore, they correspond to very large wavelengths, which do not propagate into the optical far-field because of diffraction. Zero dc component is equivalent to a vanishing time-average of the electric field, i.e., to the condition

$$\int_{-\infty}^{+\infty} E(t)\,dt = 0 \qquad (2.34)$$

for any value of the CEO phase ϕ. The $\tilde{E}(t) \propto \operatorname{sinc}(t)$ pulses (see, e.g., Fig. 2.4) we have discussed above and that we will frequently use below, do fulfill this condition for arbitrary values of ϕ. Generally, in the theory, however, one can get significant tails in the optical spectrum towards zero frequency for certain envelopes (which are well localized in time or that have steeply rising/falling edges) and values of ϕ. In this case, the electric-field envelope must not be assumed to be independent of ϕ. If one assumes a fixed envelope anyway, the light–matter interaction loses its gauge invariance [33] and unphysical results are expected.

▶ **Example 2.4.** For what pulse duration, measured in terms of the number of cycles in the pulse, do we expect unphysical results from the decomposition of the field into an envelope and a carrier-wave oscillation? Let us consider a Gaussian pulse according to $E(t) = \tilde{E}(t) \cos(\omega_0 t + \phi)$ with $\tilde{E}(t) = \tilde{E}_0 \exp(-(t/t_0)^2)$. Its temporal full width at half-maximum of the intensity profile is given by $\delta t = t_0 \, 2\sqrt{\ln\sqrt{2}} \approx 1.177 \times t_0$. Taking advantage of the mathematical identity

$$\int_{-\infty}^{+\infty} e^{-ax^2 + bx + c} \, dx = \sqrt{\frac{\pi}{a}} \, \exp\left(\frac{b^2}{4a} + c\right), \qquad (2.35)$$

the Fourier transform of the electric field, $E(\omega)$, results as

$$\begin{aligned}
E(\omega) &= \frac{1}{\sqrt{2\pi}} \int_{-\infty}^{+\infty} E(t) \, e^{+i\omega t} \, dt \\
&= \frac{\tilde{E}_0 \, t_0}{2\sqrt{2}} \left(e^{-\frac{1}{4} t_0^2 (\omega - \omega_0)^2} \, e^{-i\phi} + e^{-\frac{1}{4} t_0^2 (\omega + \omega_0)^2} \, e^{+i\phi}\right) \qquad (2.36) \\
&=: E_+(\omega) + E_-(\omega).
\end{aligned}$$

For later use, we have defined $E_+(\omega)$ and $E_-(\omega)$ as that part of the spectrum corresponding to the maximum at positive and negative frequencies ω, respectively. For $\phi = \pi/2$, the spectrum contains strictly no dc component for any value of $\omega_0 t_0$, i.e. $E(\omega = 0) = 0$. For $\phi = 0$ on the other hand, the spectrum does contain an unphysical dc component. If the pulse contains a single optical cycle (see Fig. 3.1), i.e., if $\delta t = 2\pi/\omega_0$, thus $\omega_0 t_0 \approx 0.85 \times 2\pi$, the dc component is merely 1.6×10^{-3} relative to the global maximum of the spectrum and is expected to have negligible effect. For a half-cycle pulse, $\delta t = \pi/\omega_0$, on the other hand, the dc component makes up about one third of the peak, which is definitely nonsense.

A related artifact that comes into play for such half-cycle pulses is that the spectral center of gravity of the positive-frequency part of $|E(\omega)|^2$ according to (2.37) is no longer identical to ω_0 and shifts by as much as 20% depending on the CEO phase ϕ. This shift is due to the high-frequency tail of the maximum centered at $\omega = -\omega_0$, which extends up to positive frequencies ω. For a single-cycle pulse, this shift is still negligible and we have $E(\omega) \approx E_+(\omega)$ for $\omega > 0$. We will take advantage of this approximation at several points of this book.

Other pulse envelopes than Gaussians can be slightly more "forgiving", but one should generally be very cautious with pulses shorter than one cycle of light. Usually, single-cycle pulses are unproblematic in terms of defining an envelope.   ◀

Problem 2.2. What – in principle – are the shortest *optical* pulses achievable?

2.4 A Brief Look at Phenomenological Nonlinear Optics

In order to proceed to measuring the carrier-envelope offset frequency f_ϕ in Sect. 2.6, we need a little mathematical background on nonlinear optics. Some readers may want to skip this section. We do take the opportunity, however, to precisely define what we mean by, e.g., second-harmonic generation.

In many textbooks one finds that

$$P(t) = \epsilon_0 \left(\chi^{(1)} E(t) + \chi^{(2)} E^2(t) + \chi^{(3)} E^3(t) + ... \right), \tag{2.37}$$

is the generalization of the linear optical polarization, (2.8), for large electric fields [5,22]. This is nothing but a Taylor expansion of the optical polarization P in terms of the laser electric field E. What else could one do without getting into the details ? The coefficients $\chi^{(N \neq 1)}$ are the *nonlinear optical susceptibilities* of order N, $\chi^{(1)} = \chi$ is the linear optical susceptibility. Vectors are omitted for simplicity at this point. In vectorial form, the susceptibilities would become tensors of rank N. The range of validity of (2.37) is limited. It obviously assumes an instantaneous response of $P(t)$ with respect to $E(t)$, equivalent to no or negligible frequency dependence of the $\chi^{(N)}$. This is only justified "far away" from a resonance of the material. Also, (2.37) is only really meaningful if the terms become rapidly smaller with increasing order N, i.e., if the electric field is not too large – if it is within the perturbative regime (see Problem 3.6).

The second temporal derivative of the polarization in (2.37) is the source term on the RHS of the wave equation (2.10). Consider the second-order contribution that – via the wave equation – gives rise to a second-order nonlinear contribution to the electric field $E^{(2)}(t)$. Ignoring propagation effects at this point, it is given by

$$E^{(2)}(t) \propto \chi^{(2)} \frac{\partial^2}{\partial t^2} E^2(t) \tag{2.38}$$

$$= \chi^{(2)} \frac{\partial^2}{\partial t^2} \left(\tilde{E}^2(t) \cos^2(\omega_0 t + \phi) \right)$$

$$= \chi^{(2)} \frac{\partial^2}{\partial t^2} \left(\tilde{E}^2(t) \frac{1}{2} \left(1 + \cos(2\omega_0 t + 2\phi) \right) \right).$$

The "1" in the last line reflects so-called optical rectification or the photogalvanic effect. It is not of much interest in this book because its second temporal derivative obviously vanishes. In other words: dc components do not lead to propagating electromagnetic waves. The other contribution has carrier frequency $2\omega_0$ and phase 2ϕ. We want to call it *second-harmonic generation (SHG)*.

Similarly, the third-order susceptibility leads to a third-order contribution to the electric field

$$E^{(3)}(t) \propto \chi^{(3)} \frac{\partial^2}{\partial t^2} E^3(t) \tag{2.39}$$

$$= \chi^{(3)} \frac{\partial^2}{\partial t^2} \left(\tilde{E}^3(t) \cos^3(\omega_0 t + \phi) \right)$$

$$= \chi^{(3)} \frac{\partial^2}{\partial t^2} \left(\tilde{E}^3(t) \frac{1}{4} \left(3 \cos(\omega_0 t + \phi) + \cos(3\omega_0 t + 3\phi) \right) \right),$$

which contains one contribution with carrier frequency ω_0 and phase ϕ and another one with carrier frequency $3\omega_0$ and phase 3ϕ. The latter is called *third-harmonic generation (THG)*. The term at carrier frequency ω_0 resulting from completely neglecting the temporal derivatives of the envelope $\tilde{E}^3(t)$ is called *self-phase modulation (SPM)*. For few-cycle pulses, however, the derivative of the envelope can become comparable in magnitude to that of the carrier wave. The additional term that results from accounting for a single derivative of the envelope only ($\propto \tilde{E}^2(t) \partial \tilde{E}/\partial t$) is often called *self-steepening*.

Note that the SPM is proportional to $\tilde{E}^3(t)$. If, e.g., the electric-field envelope $\tilde{E}(t)$ is a Gaussian with $\tilde{E}(t) = \tilde{E}_0 \exp(-(t/t_0)^2)$, its third power is again a Gaussian but narrower in time by a factor of $\sqrt{3}$ as we have

$$\tilde{E}^3(t) = \left(\tilde{E}_0 \, e^{-\left(\frac{t}{t_0}\right)^2} \right)^3 = \tilde{E}_0^3 \, e^{-3\left(\frac{t}{t_0}\right)^2} = \tilde{E}_0^3 \, e^{-\left(\frac{t}{t_0/\sqrt{3}}\right)^2}. \tag{2.40}$$

Correspondingly, its spectral width is larger by factor $\sqrt{3}$ – SPM broadens the spectrum. We can easily generalize this result for Gaussian pulses: The spectral width of a contribution of order N is larger than that of the fundamental by factor $\sqrt{N}$. For other shapes of the envelope, the numerical factor is different but the qualitative behavior is the same (also see Problem 2.3).

Finally, the $\chi^{(4)}$ susceptibility leads to

$$E^{(4)}(t) \propto \chi^{(4)} \frac{\partial^2}{\partial t^2} E^4(t) \tag{2.41}$$

$$= \chi^{(4)} \frac{\partial^2}{\partial t^2} \left(\tilde{E}^4(t) \cos^4(\omega_0 t + \phi) \right)$$

$$= \chi^{(4)} \frac{\partial^2}{\partial t^2} \left(\tilde{E}^4(t) \frac{1}{8} \left(3 + 4 \cos(2\omega_0 t + 2\phi) + \cos(4\omega_0 t + 4\phi) \right) \right).$$

We again get optical rectification and second-harmonic generation. The contribution with carrier frequency $4\omega_0$ and phase 4ϕ is called *fourth-harmonic generation*.

Note that our definition of a harmonic of order N is solely based on the carrier frequency and the phase. It makes no reference to the position of that contribution in the optical spectrum and does not necessarily scale with the N-th power of the incoming light (compare, e.g., SHG from a $\chi^{(2)}$ or from a $\chi^{(4)}$ process). While the carrier frequency is neither uniquely defined nor directly measurable, the phase of the N-th harmonic can be accessed via interference experiments. This aspect will be further discussed in Sect. 3.4. If a resonance comes into play, the phase of, e.g., the third harmonic can be shifted from 3ϕ to $3\phi + \delta$. The factor of "3" in front of ϕ, however, still uniquely defines third-harmonic generation. This is all we need to know in order to be able to follow the discussion on the carrier-envelope offset frequency in Sect. 2.6.

Frequently, nonlinear optics is discussed in terms of the so-called (i) *nonlinear refractive index* and (ii) the *two-photon absorption* coefficient. Both are directly related to the nonlinear optical susceptibilities. (i) Inserting the self-phase-modulation contribution of the third-order nonlinear polarization according to (2.37) on the RHS of the wave equation (2.10), bringing it to the LHS and lumping it into an effective refractive index n leads to $n = n_0 + n_2 I$. Here we have introduced the intensity I according to (2.16) and have performed a Taylor expansion with respect to I. n_0 is the linear refractive index and n_2 the nonlinear refractive index. The latter is connected to the third-order susceptibility according to $\chi^{(3)} = 4/3\, c_0 n_0^2 \epsilon_0 n_2$. (ii) The linear absorption coefficient α is proportional to the imaginary part of the complex linear refractive index. Thus, one anticipates the general intensity-dependent form $\alpha = \alpha_0 + \alpha_2 I$ in analogy to the nonlinear refractive index. To get there systematically, however, one needs to introduce a more general complex form for the nonlinear optical susceptibilities than we have done in (2.37) (where they are real). This effectively allows phase shifts between the electric field and the nonlinear polarization to be described. The two-photon absorption coefficient then turns out to be proportional to the imaginary part of the complex third-order nonlinear optical susceptibility. Readers interested in more details on these aspects of traditional nonlinear optics are referred to the excellent textbooks [5,6].

Problem 2.3. Consider the generation of the N-th harmonic via a $\chi^{(N)}$ process by a few-cycle optical pulse in the limit $N \gg 1$. Show that the shape of the N-th harmonic spectrum is approximately a Gaussian with width $\propto \sqrt{N}$ for arbitrary "well-behaved" incident pulse shapes.

2.5 Even-Harmonic Generation and Inversion Symmetry

Let us slow down for a moment and see how second-harmonic generation and even-harmonic generation in general are connected to the symmetry of the material or medium under investigation. This might avoid some confusion that could arise later in this book otherwise. Indeed, one often encounters misconceptions at this point.

Consider space inversion, i.e., we have to replace $r \rightarrow -r$. Thus, we replace $E(t) \rightarrow -E(t)$ and $P(t) \rightarrow -P(t)$ in (2.37). As $(-E(t))^2 = E^2(t)$, $(-E(t))^4 = E^4(t)$, ... it follows that $\chi^{(2)} = \chi^{(4)} = ... = 0$, while $\chi^{(1)}$, $\chi^{(3)}$, ... can be nonzero. In the previous section we have seen that second-harmonic generation requires that some $\chi^{(N)}$ with even N is nonzero. From this it follows that *second-harmonic generation is only possible in media without inversion symmetry*.

Have we really proven this statement? No, we have not! The nonlinear optical polarization P according to (2.37) depends on the electric field – but not on the magnetic field. Isn't this surprising? Wouldn't one generally expect that P is a function of both E and B? After all, the light field from a laser has an electric component and a magnetic component and the two are proportional to each other. As a result of the laser field, charged particles experience a force – the Lorentz force

$$F = q\,(E + v \times B)\,,\tag{2.42}$$

where q is the charge of the particle and v its velocity. This force gives rise to a displacement of the particle with respect to some fixed opposite charge, hence to a polarization, which depends on both electric and magnetic field of the light. The details will be discussed in Sect. 4.4[9]. It is clear, however, that v is proportional to E to the lowest order. Thus, the force and the displacement and the polarization vector are parallel to the vector product $E \times B$. For a free and nonrelativistic charge, v and E are shifted in time by 90 degrees. Having noticed all this, we have to rewrite the expression for the nonlinear optical polarization. For simplicity, we consider excitation with a plane wave with constant light intensity at this point, where $E = E_0 \cos(\omega_0 t + \phi)$ and $B = B_0 \cos(\omega_0 t + \phi)$, hence $v = v_0 \sin(\omega_0 t + \phi)$. Under these conditions, we anticipate the more general form[10]

$$P(t) = \epsilon_0 \Big(\chi^{(1)} E_0 \cos(\omega_0 t + \phi)$$
$$+ \chi^{(2)} E_0^2 \cos^2(\omega_0 t + \phi) + \chi^{(3)} E_0^3 \cos^3(\omega_0 t + \phi) + \dots$$

$$+ \chi_{\mathrm{L}}^{(2)} E_0 \times B_0 \sin(2\omega_0 t + 2\phi) + \dots \Big).\tag{2.43}$$

The last contribution $\propto E_0 \times B_0$ in this sum is new as compared to what we have in (2.37). It obeys the relation $E \perp P \perp B$. As the wavevector of light K follows $E \perp K \perp B$, we have $P \| K$ – i.e., a longitudinal polarization. This is in sharp contrast to the other terms that are transverse, i.e., which obey $P \perp K$. According to our above discussion, the term $\propto E_0 \times B_0 \sin(2\omega_0 t + 2\phi)$ corresponds to second-harmonic generation as it has carrier frequency $2\omega_0$ and phase $2\phi \pm \pi/2$. The sign depends on whether the prefactor $\chi_{\mathrm{L}}^{(2)}$ is positive or negative. Can $\chi_{\mathrm{L}}^{(2)}$ be nonzero in an inversion symmetric medium?

For space inversion we again have to replace $r \to -r$, $E \to -E$, $P \to -P$ but (watch out here!) $B \to +B$, because the magnetic field vector is an axial vector. Consequently, we have to replace $E \times B \to -E \times B$. Hence, $\chi_{\mathrm{L}}^{(2)}$ can be nonzero in an inversion symmetric medium and *second-harmonic generation can also occur in the presence of inversion symmetry*[11].

---

[9] There, we will see that the nonlinear optical polarization contains yet another strange term, namely one that, for constant light intensity, grows linearly in time t according to $P = \dots + \chi_{\mathrm{L}}^{(0)}(E_0 \times B_0)\,t$, the so-called "photon-drag" current.

[10] Remember that $\sin(\omega_0 t + \phi) \cos(\omega_0 t + \phi) = \frac{1}{2}\sin(2\omega_0 t + 2\phi)$.

[11] An optically induced magnetization M can also lead to SHG in a centrosymmetric material, for example see Ref. [34].

Even if $\chi^{(2)} = \chi^{(4)} = \ldots = \chi_{L}^{(2)} = \ldots = 0$ should hold, we generally cannot conclude that frequency doubling is only possible without inversion symmetry. Here, we define the phenomenon of *frequency doubling* by a peak or a strong contribution in the optical spectrum at the spectrometer frequency ω given by $\omega = 2\omega_0$. In Sect. 3.4 we will see that frequency doubling can even arise from the simple two-level system in the regime of extreme nonlinear optics. Moreover, it turns out that peaks at $\omega/\omega_0 = N$ with even integers N are the rule rather than the exception within the two-level system.

2.6 Principle of Measuring the Carrier-Envelope Frequency

In Sect. 2.3 we have seen that the carrier-envelope phase ϕ and the carrier-envelope offset frequency f_ϕ are important parameters of a mode-locked laser oscillator. If one wants to stabilize or control f_ϕ, one obviously first needs to be able to measure f_ϕ, equivalent to determining the pulse-to-pulse change of the CEO phase $\Delta\phi$. How can one measure ϕ of a laser pulse $E(t)$ according to (2.26)? It drops out when measuring the intensity (see Sect. 2.2), it does not affect the optical spectrum, it does not show up in usual intensity autocorrelations or field autocorrelations. Generally, in order to observe a phase, one needs to compare the unknown field with a "reference".

The idea: If one had another field that contained a phase 2ϕ rather than 1ϕ of the electric field itself, the interference of the two contributions (the beat note) would oscillate with the difference, i.e. with $(2 - 1) \times \phi = \phi$. This procedure is called *self-referencing* and was first introduced by Hänsch and coworkers [35].

Such "reference" can be generated by sending the laser electric field $E(t) = E_{\omega_0}(t) = \tilde{E}(t)\cos(\omega_0 t + \phi)$ onto a suitable nonlinear optical material, e.g., a SHG crystal. Let us consider the resulting interference of the fundamental and the second harmonic with $E_{2\omega_0}(t) \propto \tilde{E}^2(t)\cos(2\omega_0 t + 2\phi)$ in frequency space. The Fourier transforms of the cosine-terms of the different electric field contributions have maxima at positive and at negative frequencies. As in Sect. 2.3, we focus on the measurable positive frequency components (corresponding to the minus sign in the exponent). For a train of pulses, the resulting intensity from the interference can be written as

$$I_{\omega_0, 2\omega_0}(\omega) \propto \left| e^{-i\phi}\tilde{E}_{\omega_0}(\omega) + e^{-i2\phi}\tilde{E}_{2\omega_0}(\omega) \right|^2 \tag{2.44}$$
$$= \left| \tilde{E}_{\omega_0}(\omega) \right|^2 + \left| \tilde{E}_{2\omega_0}(\omega) \right|^2$$
$$+ 2\left| \tilde{E}_{\omega_0}(\omega)\,\tilde{E}_{2\omega_0}(\omega) \right| \times \cos(\phi).$$

The $\cos(\phi)$-term delivers the desired and anticipated dependence on ϕ. In order to actually observe this contribution in an experiment, at least the following two conditions have to be fulfilled.

- The amplitudes $\tilde{E}_{\omega_0}(\omega)$ and $\tilde{E}_{2\omega_0}(\omega)$ must be comparable in absolute value, otherwise the two constant, i.e., ϕ-independent, terms in (2.44) completely dominate the measured intensity $I_{\omega_0, 2\omega_0}$. This condition can generally be fulfilled at some frequency ω in the optical frequency interval $[\omega_0, 2\omega_0]$.
- The product term must exhibit appreciable absolute strength in order not to be covered by noise in the experiment.

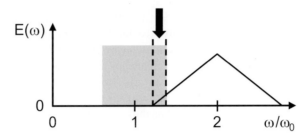

Fig. 2.5. Scheme of the laser spectrum (gray area) of a $\text{sinc}^2(t)$ pulse covering slightly more than one octave in width ($\Leftrightarrow \delta\omega/\omega_0 = 2/3$, see Problem 2.2) and its second harmonic (not to scale). In the region of overlap (see arrow) the two contributions interfere and a dependence on the CEO phase ϕ results.

It turns out that the second condition is more difficult to fulfill than the first (see Fig. 2.5). From Sect. 2.3 it is clear that a modulation according to

$$I_{\omega_0,2\omega_0} = \ldots + \ldots \cos(\phi) \tag{2.45}$$

leads to a *peak at frequency* f_ϕ *in the RF spectrum* [26, 28, 35–37]. Similarly, an interference of the third harmonic with phase 3ϕ and the fundamental with phase ϕ leads to a modulation with

$$I_{\omega_0,3\omega_0} = \ldots + \ldots \cos(2\phi), \tag{2.46}$$

equivalent to a *peak at frequency* $2f_\phi$ *in the RF spectrum*, which is most prominent somewhere in the optical frequency interval $[\omega_0, 3\omega_0]$. We will return to both types of interferences in Sects. 3.4 and 7.2.

When performing corresponding experiments, one often measures the RF power spectrum (as, e.g., in Sects. 7.1.3 or 7.2). Let us have a quick look at the details. The beat signal $I(\phi)$, i.e., $I(\phi) = I_{\omega_0,2\omega_0}(\phi)$ from (2.45) or $I(\phi) = I_{\omega_0,3\omega_0}(\phi)$ from (2.46) or some more complicated general form, can be detected by a photomultiplier tube, which delivers a proportional voltage signal $U(t)$. Assuming an integer ratio of repetition frequency and CEO frequency for simplicity at this point, i.e., $f_r/f_\phi = r$ with integer r, the signal voltage (which is illustrated in Fig. 2.6) can be written as

$$U(t) = U_0 \sum_{N_\phi=-\infty}^{+\infty} \sum_{N_r=0}^{r-1} I_{N_r} \, \delta\left(t - [N_\phi t_\phi + N_r t_r]\right), \tag{2.47}$$

with the integers N_ϕ and N_r, the abbreviation

$$I_{N_r} = I\left(\phi = N_r \frac{2\pi}{r}\right), \tag{2.48}$$

the carrier-envelope offset period $t_\phi = 1/f_\phi$, and the (unimportant) prefactor U_0. Here we have approximated the actual temporal response of the photomultiplier by a

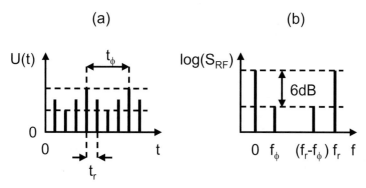

Fig. 2.6. (a) Scheme of the output voltage $U(t)$ (linear scale) of a photomultiplier detecting an optical beat signal $I(\phi) = I_{\omega_0, 2\omega_0}(\phi) = \ldots + \ldots \cos(\phi)$ according to (2.45). The ratio of repetition frequency to carrier-envelope offset frequency is $f_r/f_\phi = r = 4$ ($\Leftrightarrow \Delta\phi = \pi/2$, see Fig. 2.3). (b) Corresponding RF power spectrum $S_{RF}(f)$ on a logarithmic scale. The 50% modulation of $U(t)$ due to ϕ in (a) corresponds to peaks in the RF power spectrum at frequencies f_ϕ and $(f_r - f_\phi)$ that are $2 \times 3\,\mathrm{dB} = 6\,\mathrm{dB}$ smaller than the f_r peak and the $f = 0$ peak.

δ-function. If this voltage signal is fed into an *RF spectrum analyzer*, the *RF power spectrum* $S_{RF}(f)$ versus RF frequency f is measured. It is defined by

$$S_{RF}(f) = \left| \frac{1}{\sqrt{2\pi}} \int_{-\infty}^{+\infty} e^{i2\pi ft} U(t)\, dt \right|^2. \tag{2.49}$$

Inserting (2.47) into (2.49), the δ-functions in (2.47) select only contributions with $t = [N_\phi t_\phi + N_r t_r]$ from the integral and we obtain

$$S_{RF}(f) = \frac{U_0^2}{2\pi} \left| \sum_{N_\phi=-\infty}^{+\infty} \sum_{N_r=0}^{r-1} I_{N_r} e^{i2\pi f [N_\phi t_\phi + N_r t_r]} \right|^2 \tag{2.50}$$

$$= \frac{U_0^2}{2\pi} \left| \sum_{N_r=0}^{r-1} I_{N_r} e^{i2\pi f N_r t_r} \right|^2 \times \left| \sum_{N_\phi=-\infty}^{+\infty} e^{i2\pi f N_\phi t_\phi} \right|^2.$$

The nonvanishing values of the last sum correspond to those frequencies f, for which the terms for, e.g., N_ϕ and $(N_\phi + 1)$ add constructively, i.e., for which we have $2\pi f N_\phi t_\phi = M\, 2\pi$, thus

$$f = M/t_\phi = M\, f_\phi, \tag{2.51}$$

with integer M. This means that the RF power spectrum consists of a series of δ-peaks at integer multiples of the CEO frequency f_ϕ. The height of these peaks is given by the value of the first term in the second line of (2.50), i.e., by

$$\left| \sum_{N_r=0}^{r-1} I_{N_r} e^{i2\pi f N_r t_r} \right|^2 = \left| \sum_{N_r=0}^{r-1} I_{N_r} e^{i2\pi M\, f_\phi N_r t_r} \right|^2 = \left| \sum_{N_r=0}^{r-1} I_{N_r} e^{i2\pi M N_r/r} \right|^2. \tag{2.52}$$

In general, some of the peaks with label M may not occur because they have zero height. It is obvious that replacing M by $(M + r)$ on the RHS of (2.52) delivers the same value, i.e., the height of the peak at, e.g., frequency f_ϕ is exactly the same as the height of the peak at frequency $(f_r + f_\phi)$. In other words: All relevant information of the RF power spectrum is contained in the frequency interval $[0, f_r]$. Along the same lines, the peaks at f_ϕ and $(f_r - f_\phi)$, respectively, have the same height as well (replace M by $(r - M)$ on the RHS of (2.52)). *The mixing products* $(f_r - f_\phi)$, $(f_r - 2f_\phi)$ *or* $(f_r + f_\phi)$ *... essentially originate from the fact that the light intensity – and not the electric field itself – is subject to a Fourier transformation in an RF spectrum analyzer.* The Fourier transform of the laser electric field itself has been discussed in (2.29).

Equations (2.51) and (2.52) allow us to compute the RF power spectrum from a known beat signal $I(\phi)$. An example is given in Fig. 2.6. In a real experiment, the photomultiplier does not exhibit a δ-response. In this case, the actual voltage signal can be written as the convolution of (2.47) with the response function of the photomultiplier. In the frequency domain, this convolution translates into the product of the "ideal" result with the power spectrum of the photomultiplier response function, i.e., there is an overall decay towards large RF frequencies f. Finally, real laser systems have noise, which shows up as a pedestal in the RF power spectrum. Typically, this noise is not white noise (which would be a constant in frequency space) but is rather roughly proportional to $1/f$. This sometimes makes, e.g., the frequency interval $[2f_r, 3f_r]$ advantageous as compared to $[0, f_r]$ in the experiment – although the intervals are equivalent in theory.

Problem 2.4. Suppose you have a source of very short pulses with a fluctuating CEO phase ϕ. Can you come up with an all-optical scheme delivering (longer) pulses with a *stable* CEO phase?

Problem 2.5. Somebody provides you with a complicated laser electric field of a single pulse for CEO phase $\phi = 0$, $E(t)$, which cannot easily be decomposed into an envelope $\tilde{E}(t)$ and a carrier-wave oscillation $\cos(\omega_0 t + \phi)$. This problem occurs for practically any actual experimental pulse shape. Can you still compute the corresponding $E(t)$ for CEO phase $\phi \neq 0$?

Problem 2.6. A pulse impinges from vacuum onto a dielectric halfspace with $v_{phase} \neq v_{group} \geq 0$. How are the peak electric field and the intensity *within* the dielectric related to their counterparts in vacuum? For clarity, neglect reflection of electromagnetic energy from the interface (see Problem 2.1) and absorption.

3

The Lorentz Oscillator Model and Beyond ...

The bulk of this chapter is concerned with extreme nonlinear optics of two-level systems. Sections 3.3 to 3.6 give an overview based on exact numerical solution. Two simple approximative schemes, the "static-field approximation" and the "square-wave approximation" allow for analytical solutions and, thus, help to understand the underlying physics.

A reader familiar with two-level systems in general might want to go to Sect. 3.3 directly. For pedagogical reasons, however, we start our discussion with the classical Lorentz oscillator model, and show that its linear optical properties are identical to those of the quantum-mechanical two-level system. The latter is motivated and introduced in a way suitable for experimentalists. This is followed by a brief reminder on the "traditional" nonlinear optics of two-level systems.

3.1 Linear Optics: Revisiting the Lorentz Oscillator Model

Consider a one-dimensional harmonic oscillator with displacement x with respect to some fixed positive charge. It consists, e.g., of an electron with charge $-e$ and mass m_e subject to a Hooke spring with spring constant $\mathcal{D}$, and is excited by some laser electric field $E(t)$. Under these conditions, Newton's second law reads

$$m_e \ddot{x}(t) + \mathcal{D}x(t) = -e\,E(t)\,. \tag{3.1}$$

The dots denote the derivative with respect to time t. The optical polarization P is given by the product of the number of oscillators N_{osc} per volume V times the individual dipole moment $-e\,x$, i.e., $P = -e\,(N_{osc}/V)\,x$. Equation (3.1) is easily solved by Fourier transformation. This leads to the optical polarization P versus **spectrometer frequency** ω

$$P(\omega) = \epsilon_0 \chi(\omega) E(\omega)\,, \tag{3.2}$$

with the *linear optical susceptibility* $\chi(\omega)$ given by

$$\chi(\omega) = \frac{e^2 N_{osc}}{\epsilon_0 V m_e} \frac{1}{\Omega^2 - \omega^2}$$

$$= \frac{e^2 N_{osc}}{\epsilon_0 V m_e} \frac{1}{2\Omega} \left(\frac{1}{\Omega - \omega} + \frac{1}{\Omega + \omega} \right). \tag{3.3}$$

$\Omega = \sqrt{D/m_e}$ is the harmonic oscillator eigenfrequency or *transition frequency*. Clearly, the susceptibility (see Fig. 3.1) has poles at the two frequencies $\omega = +\Omega$ and $\omega = -\Omega$. One is tempted to argue that the negative frequency pole is irrelevant for experiments. Indeed, only the positive frequency part can be measured using a spectrometer. Thus, the negative frequency pole is not important for linear optics. We will see, however, that the corresponding contribution of the polarization at $\omega < 0$ can be the origin of unusual nonlinearities. Multiphoton absorption and carrier-wave Rabi flopping within the two-level system are just two examples of this (also, see Problem 3.4).

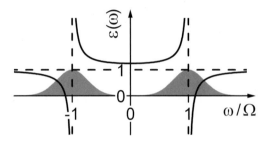

Fig. 3.1. Linear dielectric function $\epsilon(\omega) = 1 + \chi(\omega)$ versus spectrometer frequency ω according to (3.3) with transition frequency Ω. The gray areas illustrate $|E(\omega)|$ (see Example 2.4), i.e., the modulus of the Fourier transform, of a single-cycle Gaussian laser pulse $E(t) = \tilde{E}(t) \cos(\omega_0 t + \phi)$ with envelope $\tilde{E}(t) = \tilde{E}_0 \exp(-(t/t_0)^2)$ and carrier frequency $\omega_0 = \Omega$ (resonant excitation) [268].

In order to approach the mathematical form of the description in terms of two-level systems, it is instructive to rewrite the harmonic oscillator equation (3.1) slightly. Let us define the dimensionless *normalized displacement* u via

$$u = x/x_0, \tag{3.4}$$

where x_0 is some characteristic lengthscale of the problem. This leads to the equation of motion

$$\ddot{u} + \Omega^2 u = 2\Omega \Omega_R(t), \tag{3.5}$$

where we have introduced the quantity $\Omega_R(t)$ given by

$$\hbar \Omega_R(t) = d E(t). \tag{3.6}$$

$\Omega_R(t)$ obviously has the dimensions of a frequency and is called the *Rabi frequency*. d has the dimensions of a dipole and is given by $d = -e\hbar/(2m_e x_0 \Omega)$. We will see in a moment that d is the classical analogue of the dipole matrix element of the two-level system. Note that the *Rabi energy* $\hbar\Omega_R$ can be interpreted as the *electrostatic energy* of a static electric dipole with dipole moment d in the electric field of the laser.

Problem 3.1. Evaluate the group velocity $v_{group}(\omega)$ for the Lorentz oscillator model with Stokes damping γ. Can the group velocity be superluminal, i.e., $v_{group}(\omega) > c_0$? What are the implications for a short/long pulse propagating through the medium? What do you conclude from these findings in terms of the concept of the group velocity?

3.2 Two-Level Systems and Rabi Energy

The simplest nontrivial quantum-mechanical model of light–matter interaction is the two-level system [7]. For a single level, the system would simply be in its corresponding eigenstate forever and nothing would happen. Moreover, no optical dipole moment would exist, thus no coupling to the light field occurs.

The latter aspect becomes different for the two-level system. Let us, for example, consider the two lowest energy states of an electron in a box of width L with infinite potential walls[1]. The LHS wall shall be at $x = 0$, the RHS one at $x = L$. The corresponding two lowest energy solutions of the time-dependent Schrödinger equation are the (normalized) wave functions

$$\psi_1(x, t) = \sqrt{\frac{2}{L}} \sin\left(\frac{1\,\pi}{L} x\right) e^{-i\hbar^{-1}E_1 t} = \psi_1(x) e^{-i\hbar^{-1}E_1 t} \qquad (3.7)$$

$$\psi_2(x, t) = \sqrt{\frac{2}{L}} \sin\left(\frac{2\,\pi}{L} x\right) e^{-i\hbar^{-1}E_2 t} = \psi_2(x) e^{-i\hbar^{-1}E_2 t}. \qquad (3.8)$$

$E_1 = \hbar^2(1\pi/L)^2/(2m_e)$ and $E_2 = \hbar^2(2\pi/L)^2/(2m_e)$ are the corresponding eigenenergies, $\psi_1(x)$ and $\psi_2(x)$ the corresponding eigenfunctions of the stationary Schrödinger equation. Note that the charge density corresponding to state 1, which is proportional to $-e\,|\psi_1(x, t)|^2$, is constant in time. Thus, light neither is emitted nor absorbed. The same argument holds for state 2. It is only for a superposition state such as, e.g., the normalized wave function

$$\psi(x, t) = \frac{1}{\sqrt{2}}\left(\psi_1(x, t) + \psi_2(x, t)\right) \qquad (3.9)$$

that the charge density ρ with

---

[1] While the potential well is just a toy model at this point, it actually describes intersubband transitions in semiconductor quantum wells rather well [40].

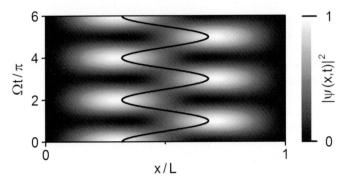

Fig. 3.2. Grayscale image of the probability density $|\psi(x,t)|^2$ (linear scale) of an electron in a two-level system versus x coordinate and time t according to (3.9). The solid black curve reveals the expectation value $\langle x \rangle$ of the electron displacement with $\langle x \rangle/L = \frac{1}{2} - \frac{16}{9\pi^2}\cos(\Omega t)$ (also see Problem 3.2). The transition frequency Ω is given by $\hbar\Omega = E_2 - E_1$ [268].

$$\rho(x,t) = -e\,|\psi(x,t)|^2 = -e\,\frac{1}{2}\,|\psi_1(x,t) + \psi_2(x,t)|^2 \qquad (3.10)$$

oscillates in time. Figure 3.2 illustrates this oscillation. One gets a harmonic oscillation along x with the *transition frequency* Ω given by

$$\hbar\Omega = E_2 - E_1 . \qquad (3.11)$$

The analogy to the classical harmonic oscillator discussed in the previous section is more than obvious.

What is the minimum set of variables we need in order to describe the dynamics of the two-level system? We clearly want to know the occupancy of the levels 1 and 2. Let us call them f_1 and f_2. As the total occupation of electrons in the two levels is one, i.e., $f_1 + f_2 = 1$, we can introduce the *inversion* w given by $w = f_2 - f_1$. If we know w, f_1 and f_2 follow immediately. Furthermore, we have seen that the oscillating dipole is intimately related to the superposition state. To make an optical transition from level 1 to level 2, one first needs to excite the superposition state – as only this superposition state couples to the light. Thus, we anticipate that the mathematical description must not only contain the inversion w but also describe the superposition state. As we could add a complex prefactor in front of ψ_2 in (3.10), we need two additional real quantities, encoding amplitude and phase of the superposition state. Altogether, we expect that the mathematical description of the two-level system will invoke three real quantities forming the Bloch vector.

Let us derive the equation of motion of this Bloch vector. The coupling of laser light and matter is often well described within the so-called *dipole approximation*. Here one takes advantage of the fact that the wavelength of light is typically much larger than the size of a dipole or larger than the lattice constant in a solid. This means that the frequency of light can be considered as small, hence, the rules of electrostatics apply. In electrostatics, the interaction energy of a dipole d in a static electric field E

is given by $-dE$. In quantum mechanics, d becomes the *dipole matrix element*, i.e.,

$$d = \int_{-\infty}^{+\infty} \psi_2^*(x)(-e\,x)\,\psi_1(x)\,dx\,, \tag{3.12}$$

where $\psi_1(x)$ and $\psi_2(x)$ are the (normalized) solutions of the stationary Schrödinger equation. In what follows, we will assume that d is real. In addition to this interaction energy we have the energy of the two-level system itself, which is given by the occupation of level 1 times its energy E_1 plus the occupation of level 2 times energy E_2.

In "second quantization" this corresponds to the model Hamiltonian

$$\mathcal{H} = E_1\,c_1^\dagger c_1 + E_2\,c_2^\dagger c_2 - d\,E(\boldsymbol{r}, t)\left(c_1^\dagger c_2 + c_2^\dagger c_1\right)\,. \tag{3.13}$$

The creation $c^\dagger$ and annihilation c operators create and annihilate electrons in states number 1 or 2, respectively. $c_1^\dagger c_1$ counts the occupation of level 1, $c_2^\dagger c_2$ that of level 2. The expectation values $\langle c_1^\dagger c_1\rangle = f_1$ and $\langle c_2^\dagger c_2\rangle = f_2$ are the occupation numbers of levels 1 and 2. $c_2^\dagger c_1$ annihilates an electron in level 1 and creates one in level 2, $c_1^\dagger c_2$ promotes an electron from level 2 to level 1. This is clearly related to optical transitions, and hence related to the optical polarization. We define the *Bloch vector* $(u, v, w)^{\mathrm{T}}$ via

$$\begin{pmatrix} u \\ v \\ w \end{pmatrix} := \begin{pmatrix} \langle c_1^\dagger c_2\rangle + \langle c_2^\dagger c_1\rangle \\ -i\,(\langle c_1^\dagger c_2\rangle - \langle c_2^\dagger c_1\rangle) \\ \langle c_2^\dagger c_2\rangle - \langle c_1^\dagger c_1\rangle \end{pmatrix}\,. \tag{3.14}$$

It is straightforward to calculate the equation of motion of the Bloch vector by evaluating the operators in the expectation values via the Heisenberg equation of motion for an arbitrary operator $\mathcal{O}$ according to

$$-i\hbar\frac{\partial}{\partial t}\mathcal{O} = [\mathcal{H}, \mathcal{O}]\,. \tag{3.15}$$

Employing the usual anticommutation rules for fermions, i.e.,

$$[c_1, c_1^\dagger]_+ = 1\,, \qquad [c_2, c_2^\dagger]_+ = 1\,, \tag{3.16}$$

and that all other anticommutators are zero, leads us to the well-known *Bloch equations* [51, 52] in matrix form

$$\begin{pmatrix} \dot{u} \\ \dot{v} \\ \dot{w} \end{pmatrix} = \begin{pmatrix} 0 & +\Omega & 0 \\ -\Omega & 0 & -2\,\Omega_R(t) \\ 0 & +2\,\Omega_R(t) & 0 \end{pmatrix} \begin{pmatrix} u \\ v \\ w \end{pmatrix}\,. \tag{3.17}$$

Here we have introduced the *Rabi frequency* $\Omega_R(t)$ according to

$$\hbar \Omega_R(t) = d\, E(t) \tag{3.18}$$

with the laser electric field

$$E(t) = \tilde{E}(t) \cos(\omega_0 t + \phi). \tag{3.19}$$

Note that the definition of the Rabi energy, (3.18), is identical to that introduced for the classical Lorentz oscillator, (3.6), except for a different expression for d in the classical case. Finally, considering that we have a number N_{2LS} of two-level systems per volume V, the *optical polarization* is given by

$$P(\mathbf{r}, t) = \frac{N_{2LS}}{V}\, d\, u. \tag{3.20}$$

The Bloch vector parametrically depends on $\mathbf{r}$ (we suppress this dependence for clarity). The (3×3) matrix in the equation of motion (3.17) of the Bloch vector only leads to rotations: The Bloch vector rotates with the transition frequency Ω in the uv-plane and with $2\Omega_R(t)$ in the vw-plane. This looks quite simple. Note, however, that the Rabi frequency $\Omega_R(t)$ itself oscillates with the carrier frequency of light ω_0 and periodically changes sign. *It is this aspect that makes the behavior so rich.* Together with ω we have the four important frequencies of the problem

- Spectrometer frequency ω
- Transition frequency Ω
- Rabi frequency $\Omega_R(t)$ with peak Ω_R
- Carrier frequency of light ω_0

Let us have a brief look at the limit of linear optics. It is defined by only a negligible amount of electrons being promoted from level 1 into level 2, hence we can approximate the inversion as $w = -1$. Inserting $w = -1$ into the first line of (3.17), taking its temporal derivative and introducing the second line for $\dot{v}$, we recover (3.5). We conclude that *we can interpret the component u of the Bloch vector as the analogue of the normalized displacement of the Lorentz oscillator.* Indeed, as u is proportional to the optical polarization P, u contains all the information on the optical properties of the material.

Problem 3.2. Compute the dipole matrix element d corresponding to an optical transition between states 1 and 2 of the simple model of an electron in a box with infinite potential walls.

Problem 3.3. Consider resonant excitation of a two-level system in the *incoherent limit*. Specifically, compute the steady-state inversion, w, of the two-level system via the optical Bloch equations (3.17) for the Bloch vector $(u, v, w)^T$ and account for dephasing according to $\dot{u} = \ldots - u/T_2$ and $\dot{v} = \ldots - v/T_2$. T_2 is called the dephasing time or transverse relaxation time. The latter notion originates from nuclear magnetic resonance (NMR) [51, 52], where the components u and v of the Bloch vector correspond to the real space x and y-components of the magnetization. x and y are perpendicular (transverse) to the static magnetic field, which is usually oriented along the z-direction.

3.3 Carrier-Wave Rabi Flopping

Let us now jump right into nonlinear optics in the following fashion: First, we discuss the simplest case, namely resonant excitation of a two-level system ($\Omega/\omega_0 = 1$) for Rabi frequencies approaching the carrier frequency of light, i.e., $\Omega_R/\omega_0 \approx 1$. We will encounter some famous examples from traditional nonlinear optics on the way and arrive at carrier-wave Rabi flopping. For off-resonant excitation such as, e.g., $\Omega/\omega_0 = 2$, the behavior is quite different and we will, for example, encounter "third-harmonic generation in the disguise of second-harmonic generation". Only for the limit $\Omega/\omega_0 \gg 1$, are our expectations from phenomenological nonlinear optics (Sect. 2.4) recovered. For yet larger Rabi energies, $\Omega_R/\omega_0 > 1$ or even $\Omega_R/\omega_0 \gg 1$, high harmonics are generated. We will see that spectral peaks at even integer multiples of the laser carrier frequency are generally connected with points of commensurability of the three involved frequencies: ω_0, Ω and Ω_R.

If a two-level system is excited by a resonant light field ($\Omega/\omega_0 = 1$), electrons absorb photons that pump them from the ground state into the excited state (see Fig. 3.3). It is sometimes stated that one cannot reach inversion by optical pumping in a two-level system. This statement is, however, only true in the incoherent steady-state case, where one can only reach transparency indeed, i.e., 50% of the electrons are in the ground state, 50% are in the excited state, the inversion is $w = 0$. In contrast to this, if the system remains fully coherent in the quantum-mechanical sense – as discussed in the previous section – complete inversion can be reached. If the light field remains switched on, stimulated emission brings the electrons back into the ground state. This oscillation of the inversion is known as Rabi oscillation or *Rabi flopping* [8, 22, 53]. Note that one must not interpret the dots in Fig. 3.3 as the electrons. Remember that the electrons are in a superposition state of state 1 and state 2. One can, however, interpret the dots as a measure of the occupation of levels 1 and 2, respectively.

Fig. 3.3. Scheme of a Rabi oscillation in a two-level system versus time t. The lower horizontal lines represent the ground state, the upper lines the excited state. The dots symbolize the electron occupation numbers [268].

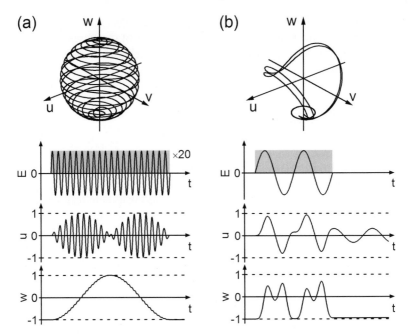

Fig. 3.4. Illustration of **(a)** conventional Rabi flopping and **(b)** carrier-wave Rabi flopping. In (a), the exciting box-shaped electric field pulse $E(t)$ contains $N = 20$ cycles of light and is weak, i.e., the peak Rabi frequency is given by $\Omega_R/\omega_0 = 1/20 \Rightarrow \tilde{\Theta} = 2\pi$. As a result, the inversion w (illustrated by the insets) nearly reproduces the classical result of Rabi according to $w(t) = -\cos(\tilde{\Omega}_R t)$. A closer inspection does, however, reveal a small superimposed, staircase-like structure (the Bloch–Siegert oscillation) which is due to deviations from the rotating-wave approximation used by Rabi. These minor structures become large and prominent qualitative changes in (b), where we depict the behavior for a $N = 2$ cycle long, intense pulse with $\Omega_R/\omega_0 = 1 \Rightarrow \tilde{\Theta} = 4\pi$ (shown on a stretched timescale). Here, both the inversion $w(t)$ as well as the optical polarization $P(t) \propto u(t)$ exhibit significant dynamics on the timescale of one cycle of light – in contrast to conventional Rabi flopping (a). For example, for a photon energy of $\hbar\omega_0 = 1.5\,eV$, a cycle of light has a period of 2.8 fs and the pulse envelope (gray area) in (b) corresponds to a duration of 5.6 fs. Figure reprinted from Ref. [54] by permission of O. D. Mücke.

An actual solution of the Bloch equations in the regime of traditional nonlinear optics is depicted in Fig. 3.4(a). The evolution of the Bloch vector is shown on the top LHS of Fig. 3.4. The inversion versus time (Fig. 3.4(a)) exhibits a slow oscillatory behavior with a rapid and very weak oscillation superimposed. The latter is sometimes referred to as the *Bloch–Siegert oscillation* [55]. In the original approach of nobel prize winner I. I. Rabi[2], he used the so-called rotating-wave approximation in which only resonant contributions are accounted for (he left aside the pole at negative frequencies discussed in Sect. 2.3). This neglects the small ripples and leads to the simple closed

---

[2] Actually, Rabi discussed a magnetic dipole in a rotating magnetic field.

form for the inversion

$$w(t) = -\cos(\tilde{\Omega}_R t),\qquad(3.21)$$

where we have assumed that the system is in its ground state at time $t = 0$. The quantity $\tilde{\Omega}_R$ is the *envelope Rabi frequency* and is given by

$$\hbar\tilde{\Omega}_R(t) = d\tilde{E}(t).\qquad(3.22)$$

Note that the envelope Rabi frequency no longer oscillates with the carrier frequency of light. The pulse depicted in Fig. 3.4(a) is called a 2π-pulse as it leads to one complete Rabi oscillation within the pulse. One also says that the (envelope) pulse area is 2π. Similarly, a $\tilde{\Theta}$-pulse has an *envelope pulse area* of

$$\tilde{\Theta} = \int_{-\infty}^{+\infty} \tilde{\Omega}_R(t)\,dt.\qquad(3.23)$$

The modulation of the optical polarization with the Rabi frequency in the time domain clearly corresponds to sidebands in the Fourier domain. The smaller the period of the Rabi oscillation, the larger the separation of the sidebands in the optical spectrum. Indeed, Mollow [56–59] showed (within the rotating-wave approximation as Rabi) that, for constant light intensity, one gets a triplet, the famous *Mollow triplet*, consisting of three lines at spectrometer frequencies $\omega = \Omega$, $\omega = \Omega + \tilde{\Omega}_R$ and $\omega = \Omega - \tilde{\Omega}_R$. This triplet can also be seen in the exact numerical solutions of the Bloch equations shown in Fig. 3.5 (around $\omega/\omega_0 = 1$ on the horizontal axis and for $\Omega_R/\omega_0 \ll 1$ on the vertical axis). The Mollow triplet is simply the resonant analogue of self-phase modulation for off-resonant excitation (see Sect. 2.4). Both broaden the incoming laser spectrum. The shape of the resulting spectrum is different, however. We will see in the next section that there is a continuous transition between the Mollow triplet and self-phase modulation when changing Ω/ω_0.

▶ **Example 3.1.** For an electric field with $\tilde{E}_0 = 4 \times 10^9$ V/m and for GaAs parameters ($d = 0.5\,e$ nm, with the elementary charge $e = 1.6021 \times 10^{-19}$ A s), we obtain $\hbar\Omega_R = 2\,eV$, which is comparable to the photon energy of $\hbar\omega_0 = 1.42\,eV$ corresponding to the GaAs bandgap (see Sect. 7.1.1).   ◀

The notion *carrier-wave Rabi flopping* has been introduced by Hughes [60, 61] in 1998 (also see Refs. [62–66]) and refers to a situation in which the Rabi frequency becomes comparable to the laser carrier frequency. An intuitive understanding can be obtained from the top of Fig. 3.4. For conventional Rabi flopping, (a), the Bloch vector rapidly orbits around the equatorial plane with the transition frequency and slowly rotates from the south pole to the north pole and back down again with the Rabi frequency. When the two frequencies become comparable, (b), the trace of the

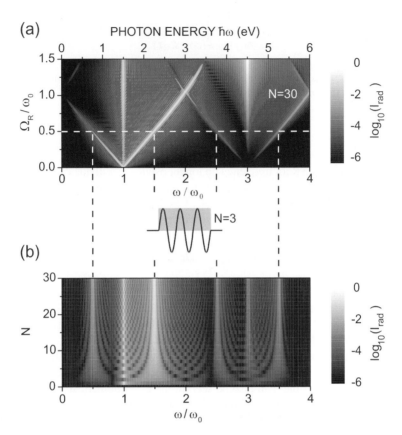

Fig. 3.5. Grayscale plots of the radiated light intensity I_{rad} (i.e., normalized square modulus of the Fourier transform of the second temporal derivative of the optical polarization $P(t) \propto u(t)$ from the Bloch equations) versus spectrometer frequency ω. **(a)** Position of the peaks of the fundamental and the third-harmonic Mollow triplet versus peak Rabi frequency Ω_R in units of the laser carrier frequency ω_0 for $\Omega/\omega_0 = 1$ with $\hbar\omega_0 = 1.5\,eV$ and for a $N = 30$-cycle long box-shaped optical pulse (see center inset for $N = 3$). **(b)** Dependence on the integer number of cycles $N = 1, 2, ...30$ in the pulse for fixed $\Omega_R/\omega_0 = 0.5$. Note the occurrence of additional side maxima for few-cycle pulses [269].

Bloch vector looks rather complicated and, as a result, the optical polarization $\propto u(t)$ no longer oscillates sinusoidally (see Fig. 3.4(b)). This clearly leads to harmonics in the Fourier domain. This can be seen in Fig. 3.5(a) around $\omega/\omega_0 = 3$ on the horizontal axis and starting above $\Omega_R/\omega_0 = 0.1$ on the vertical axis where another triplet occurs. We will call it the (third-harmonic) *carrier-wave Mollow triplet*. It consists of three peaks, approximately at spectrometer frequencies $\omega = 3\Omega$, $\omega = 3\Omega + \Omega_R$ and $\omega = 3\Omega - \Omega_R$ (we will see in Sect. 3.6 that deviations from this form occur for yet larger Rabi frequencies). For a laser intensity corresponding to $\Omega_R/\omega_0 = 1$, the high-frequency peak of the fundamental Mollow triplet and the low-frequency peak of the third-harmonic carrier-wave Mollow triplet meet at spectrometer frequency $\omega/\omega_0 = 2$ – frequency doubling in an inversion symmetric medium. Following our discussion in Sect. 2.6 on measuring the carrier-envelope offset phase, we expect that the resulting interference of the fundamental with phase ϕ and the third-harmonic with phase 3ϕ leads to a beating with difference phase $3\phi - \phi = 2\phi$. As ϕ changes with the carrier-envelope offset frequency f_ϕ, we anticipate a contribution at frequency $2f_\phi$ in the radio-frequency spectrum. We will come back to corresponding experiments in Sect. 7.2.

Another consequence of carrier-wave Rabi flopping is that the inversion w also behaves unusually. A $\tilde{\Theta} = 2\pi$ envelope pulse area in Fig. 3.5(a) indeed leads to one complete Rabi flop, i.e., after the pulse the system comes back to its ground state with $w = -1$. This statement is true for any carrier-envelope phase ϕ. For (b), this is not the case (see $w(t)$ and especially $u(t)$ after the pulse), even though the envelope pulse area of $\tilde{\Theta} = 4\pi$ suggests that. We conclude that the concept of the envelope pulse area loses its immediate meaning in the regime of carrier-wave Rabi flopping. As a consequence of this, the so-called *area theorem* of nonlinear optics also no longer applies as it is directly based on the envelope pulse area. The area theorem refers to the propagation of an optical pulse through an inhomogeneously broadened ensemble of two-level systems along the z-direction within the rotating-wave approximation and within the slowly varying envelope approximation (see Sect. 6.2). It makes the following statement about the envelope pulse area

$$\frac{d\tilde{\Theta}}{dz} = -\frac{\alpha}{2} \sin(\tilde{\Theta}(z)), \qquad (3.24)$$

which means that pulses with envelope areas $\tilde{\Theta} = N\, 2\pi$ with integer N propagate through the medium without changing their area. The pulse shape can still change. Solutions with $\tilde{\Theta} = N\,\pi$ and odd integer N turn out to be unstable. In the linear optical limit, the area theorem reduces to Beer's law describing an exponential decay along z with the intensity absorption coefficient α, because there $\sin(\tilde{\Theta}) \approx \tilde{\Theta}$.

Let us just mention that, as a result of the failure of the envelope-area concept, the inversion after the pulse, i.e., $w(t \to \infty)$, exhibits a dependence on the carrier-envelope offset phase ϕ of the pulse. This inversion determines the photocurrent in a hypothetical photodetector based on two-level system optical transitions and one would directly obtain a photocurrent depending on ϕ. We come back to this aspect in Sect. 3.5.

Problem 3.4. We have seen in Fig. 3.4 that, for resonant excitation, the two-level system inversion $w(t)$ exhibits a component that oscillates with twice the carrier frequency of light (see, e.g., the staircase-like structure in (a)). Give an intuitive explanation (no calculation!) based on Fig. 3.1.

3.4 Frequency Doubling with Inversion Symmetry

Let us come back to our statement of the previous section that the Mollow triplet is simply the resonant analogue of self-phase modulation. Figure 3.6(c) exhibits a clear Mollow splitting around the fundamental, i.e., around $\omega/\omega_0 = 1$ on the horizontal axis for $\Omega/\omega_0 = 1$ on the vertical axis. Going upwards towards larger values of Ω/ω_0 reduces the contribution on the left and on the right of the laser until only tails remain, e.g., for $\Omega/\omega_0 = 2$. This example is indeed highlighted by the white curve, which is plotted on a linear scale. Where exactly this transition from resonant to off-resonant occurs is determined by the width of the laser spectrum (see gray areas on the RHS). In addition, not surprisingly, a peak at $\omega/\omega_0 = 3$ also occurs, which is simply third-harmonic generation. In Figs. 3.6(a) – (c), this white curve is qualitatively well described by the phenomenological off-resonant perturbative self-phase modulation and third-harmonic generation discussed in Sect. 2.4. For yet larger Rabi frequencies, however, the behavior changes here as well and a peak evolves at a spectrometer frequency $\omega/\omega_0 = 2$. This peak becomes the most prominent feature in the optical spectrum for (d) where $\Omega_R/\omega_0 = 2$. Let us have a closer look at this frequency-doubling contribution in Fig. 3.7, where $\Omega_R/\omega_0 = 0.76$. Fixing the transition frequency to $\Omega/\omega_0 = 2$ in (b), allows us to study the dependence on the carrier-envelope offset phase ϕ. On the LHS and RHS of the spectrometer frequency $\omega/\omega_0 = 2$, a modulation versus ϕ with a period of π can be seen. We already suspect that both contributions arise as a result of the interference of the fundamental and the third-harmonic, giving rise to a difference in phase of 2ϕ, hence to a beat period of π (rather than 2π).

To better understand this unusual frequency-doubling contribution, we fix the CEO phase to $\phi = 0$ in Fig. 3.7(c) and vary the pulse duration t_{FWHM} from 5 fs to 10 fs. Under these conditions, the frequency-doubling peak disappears at around pulse durations of 8 fs and we conclude that this unusual contribution requires both short and intense laser pulses. Intuitively, this behavior arises from the enormous broadening of the peak centered around the third harmonic, which, at some point, exhibits a significant overlap with the resonance at $\omega/\omega_0 = 2 = \Omega/\omega_0$. This can indeed nicely be seen in Fig. 3.7(c). The resonance picks up this contribution and amplifies it by orders of magnitude. Short pulses quite obviously broaden the spectrum. Moreover, at large Rabi frequencies, higher and higher order contributions come into play (also see Fig. 3.8). For example, for a Gaussian laser pulse, the spectral width of the N-th order contribution scales as $\sqrt{N}$ (as discussed in Sect. 2.4). The combination of the two aspects – short pulses and high intensities – makes this effect appreciable in strength.

As the two-level system has inversion symmetry, the frequency doubling cannot be due to a $\chi^{(2)}$ effect. It is rather due to third-harmonic generation (THG) that looks like second-harmonic generation (SHG) at first sight, thus, it has been named *"THG in the disguise of SHG"*. We will describe corresponding experiments on semiconductors in Sect. 7.2.

Problem 3.5. Consider a sinc^2-pulse with $\hbar\omega_0 = 1.5\,eV$ exciting a narrow two-level resonance at $\Omega/\omega_0 = 2$. What is the maximum pulse duration t_{FWHM} that leads to

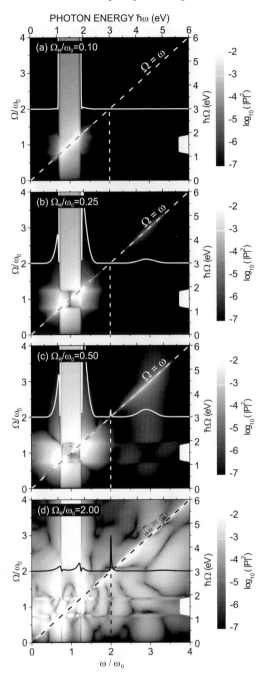

Fig. 3.6. Grayscale image of $|P|^2$ (normalized) versus ω and versus Ω, both in units of ω_0 with $\hbar\omega_0 = 1.5\,eV$. The peak Rabi frequency Ω_R of the exciting $\mathrm{sinc}^2(t)$ pulses with duration $t_{FWHM} = 5\,\mathrm{fs}$ and $\phi = 0$ is parameter. (a) $\Omega_R/\omega_0 = 0.10$, (b) $\Omega_R/\omega_0 = 0.25$, (c) $\Omega_R/\omega_0 = 0.50$, and (d) $\Omega_R/\omega_0 = 2.0$ [269].

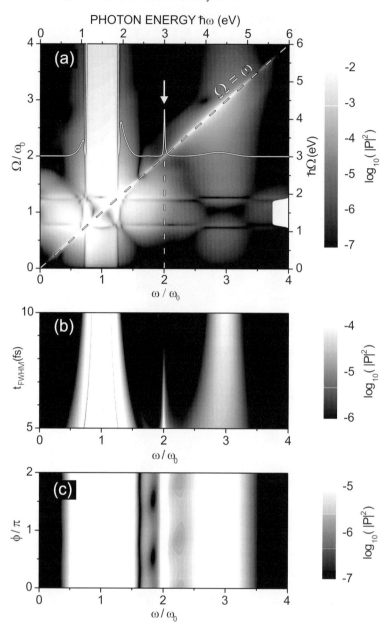

Fig. 3.7. As Fig. 3.6 but for a Rabi frequency of $\Omega_R/\omega_0 = 0.76$. **(a)** $|P(\omega)|^2$ versus transition frequency Ω for a fixed CEO phase $\phi = 0$ and $t_{FWHM} = 5$ fs. The white curve is a cut through the data at $\Omega/\omega_0 = 2$ (linear scale). The laser pulse spectrum is shown as the gray area on the RHS. **(b)** $|P(\omega)|^2$ versus pulse duration t_{FWHM} for fixed $\Omega/\omega_0 = 2$ and $\phi = 0$. **(c)** $|P(\omega)|^2$ versus ϕ for fixed $\Omega/\omega_0 = 2$ and $t_{FWHM} = 5$ fs. Reprinted with permission from T. Tritschler et al., Phys. Rev. Lett. **90**, 217404 (2003) [67]. Copyright (2003) by the American Physical Society.

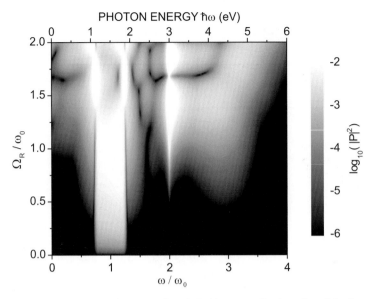

Fig. 3.8. As Fig. 3.7, but as a function of peak Rabi energy Ω_R in units of the laser carrier frequency ω_0 for fixed $\Omega/\omega_0 = 2$, $\phi = 0$ and $t_{\text{FWHM}} = 5\,\text{fs}$. Around $\Omega_R/\omega_0 \approx 1.7$ on the vertical axis, a Rabi flop is completed even though the excitation is off-resonant for one-photon absorption [269].

third-harmonic generation in the disguise of second-harmonic generation in the $\chi^{(3)}$ limit?

Problem 3.6. Consider *far off-resonant excitation* ($\Omega/\omega_0 \gg 1$) of a two-level system in the *perturbative limit* ($\Omega_R/\Omega \ll 1$). Try to recover the form for the optical polarization $P(t) = \epsilon_0(\chi^{(1)}E(t) + \chi^{(2)}E^2(t) + \chi^{(3)}E^3(t) + ...)$ according to (2.37). What can you say about the third-order susceptibility? Hint: Introduce transverse and longitudinal damping along the lines of Problem 3.3.

3.5 Quantum Interference of Multiphoton Absorption

So far, we have concentrated on the behavior of the optical polarization, i.e., on the component u of the Bloch vector. In this section we focus our attention on the inversion w and its dependence on the carrier-envelope offset phase ϕ. The inversion versus time on the timescale of a cycle of light is not really an observable – at least by today's standards. However, what one can measure is the inversion remaining after the optical pulse, i.e., $w(t \to \infty)$. This can be accomplished by sweeping out the electrons in the upper state in the static electric field of a photodetector and detecting the resulting photocurrent, or, alternatively, by an additional optical probe pulse at times well after the excitation pulse. This weak probe pulse would see a reduced absorption or even experience stimulated emission depending on the value of $w(t \to \infty)$. Let

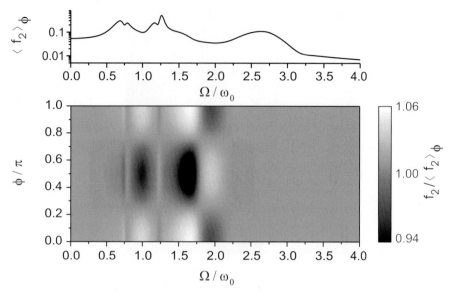

Fig. 3.9. A two-level system with transition frequency Ω is excited with a $t_{FWHM} = 5$ fs sinc^2-shaped optical pulse with carrier photon energy $\hbar\omega_0 = 1.5\,eV$ and CEO phase ϕ. The peak Rabi frequency is given by $\Omega_R/\omega_0 = 1$, the dephasing time is $T_2 = 50$ fs and $T_1 = \infty$. The resulting occupation of the upper state long after the pulse, i.e., $f_2 = f_2(t \rightarrow \infty) = \frac{1}{2}(w(t \rightarrow \infty)+1)$, normalized by the average of f_2 with respect to ϕ, $\langle f_2 \rangle_\phi$, is plotted on a grayscale versus Ω/ω_0 and ϕ. Note the π periodicity versus ϕ. The upper plot shows the average $\langle f_2 \rangle_\phi$ on a logarithmic scale. The maximum relative modulation of $f_2/\langle f_2 \rangle_\phi$ is on the order of a few per cent. Reducing the peak Rabi frequency from $\Omega_R/\omega_0 = 1$ down to $\Omega_R/\omega_0 = 0.5$ leads to a rather similar qualitative scenario (Fig. 3.10), however, with a maximum relative modulation enormously reduced to about $\pm 10^{-4}$. Also compare with Fig. 3.4 [270].

us remind ourselves that within the concept of the envelope pulse area (traditional nonlinear optics), strictly no dependence of $w(t \rightarrow \infty)$ on ϕ is expected at all. There, $w(t \rightarrow \infty)$ solely depends on the envelope pulse area $\tilde{\Theta}$.

We have already seen in Sect. 2.6 as well as in the preceding section that a dependence on ϕ arises from the interference of different pathways. In the language of quantum optics, two examples of pathways from the ground state towards the excited state of the two-level system are one-photon absorption and two-photon absorption. The one-photon process has phase ϕ, the two-photon absorption has phase 3ϕ because it can be expressed as a $\chi^{(3)}$ process. Thus, the interference of one-photon and two-photon absorption leads to a dependence on the CEO phase ϕ with period π. This statement is true for a single sufficiently intense few-cycle optical pulse, as illustrated in Figs. 3.9 and 3.10. Here, very loosely speaking, one-photon absorption of photons from the high-energy end of the laser spectrum interferes with two-photon absorption of photons from the low-energy end. Note that the dependence on ϕ changes phase depending on Ω/ω_0. For example, for $\Omega/\omega_0 = 2.0$ one has a maximum at $\phi = \pi/2$ and minima at $\phi = 0, \pi$, whereas for $\Omega/\omega_0 = 1.7$ the situation is reversed. Thus,

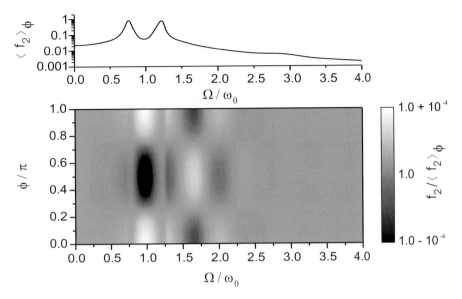

Fig. 3.10. As Fig. 3.9, but for $\Omega_R/\omega_0 = 0.5$ instead of $\Omega_R/\omega_0 = 1.0$ [270].

integration over a certain range of transition frequencies Ω/ω_0 (as one would have for band-to-band transitions in a semiconductor photodetector, where $E_g \leq \hbar\Omega < E_{max}$ with the bandgap energy E_g and some effective cutoff energy E_{max}) reduces the degree of modulation with respect to the CEO phase ϕ.

In order to understand and discuss the complex behavior in Fig. 3.9 in some more detail, we consider two simple and instructive examples that can be treated analytically: Two-color excitation and box-shaped pulses.

Two-color excitation

Suppose that the incident electric field is given by the sum of two monochromatic contributions, i.e., by

$$E(t) = \tilde{E}_1 \cos(\omega_1 t + \varphi) + \tilde{E}_2 \cos(\omega_2 t + \varphi), \tag{3.25}$$

with phase φ. Note that φ does *not* just shift the time axis but rather leads to inequivalent electric fields for $\omega_1 \neq \omega_2$ and nonzero amplitudes $\tilde{E}_1$ and $\tilde{E}_2$. Loosely speaking, the two frequencies ω_1 and ω_2 can be viewed as two components out of the broad spectrum of a short laser pulse[3]. Alternatively, we can consider the special case $\tilde{E}_1 =$

---

[3] Excitation of semiconductors with a field of the form $E(t) = \tilde{E}_1 \cos(\omega_1 t + \varphi_1) + \tilde{E}_2 \cos(\omega_2 t + \varphi_2)$ with $\omega_2 = 2\omega_1$ has recently been discussed in a series of theoretical and experimental papers, which explicitly exploit the lack of inversion symmetry. This allows for the independent control of optically induced charge currents [68–70] as well as of spin currents [71–73] via variation of the *difference phase* $(\varphi_2 - 2\varphi_1)$ and the light polarizations with respect to the crystallographic axes. The underlying physics is related to but yet distinct from that of this section (note that we discuss the case $\varphi_2 = \varphi_1 = \varphi$). Also see Ref. [74] for application to measuring the carrier-envelope offset frequency.

$\tilde{E}_2 = \tilde{E}_0$. Using the mathematical identity $\cos(a) + \cos(b) = 2\cos(\frac{a+b}{2})\cos(\frac{a-b}{2})$, the electric field according to (3.25) can be rewritten as

$$E(t) = \tilde{E}_0 \cos(\omega_1 t + \varphi) + \tilde{E}_0 \cos(\omega_2 t + \varphi)$$
$$= \tilde{E}(t)\cos(\omega_0 t + \phi), \qquad (3.26)$$

where we have introduced the effective envelope $\tilde{E}(t) = 2\tilde{E}_0 \cos(\frac{\omega_1-\omega_2}{2}t)$ (the difference-frequency oscillation – a pulse train) and the effective carrier frequency $\omega_0 = \frac{\omega_1+\omega_2}{2}$ (the sum-frequency oscillation). The CEO phase ϕ of the effective pulse centered around $t = 0$ is given by

$$\phi = \varphi. \qquad (3.27)$$

Solving the two-level system Bloch equations (3.17) in a perturbational approach in terms of orders of the incident electric field starting from the ground state, the inversion is $w(t) = -1$ to zeroth order in the field. The first order of the components u and v of the Bloch vector clearly comprise the frequencies ω_1 and ω_2, because they are driven directly by $-2\Omega_R(t)w(t) \approx +2\Omega_R(t)$ (see (3.17)). The lowest nontrivial order of the inversion w is the *second order* (one-photon absorption) because w is driven by the product $2\Omega_R(t)v(t)$. This second order contains the frequency components $\omega_1 \pm \omega_2$, $2\omega_1$, $2\omega_2$, $\omega_1 - \omega_1$ and $\omega_2 - \omega_2$. *Only the latter two have a nonvanishing cycle-average $\langle...\rangle$ and lead to a net increase of the upper-state population* $f_2 = (w+1)/2$ after some time, provided that the transverse damping is finite (otherwise $\langle\Omega_R(t)v(t)\rangle \propto \langle\cos(\omega_{1,2}t)\sin(\omega_{1,2}t)\rangle = 0$). All other contributions merely lead to an oscillatory component (also see $w(t)$ in Fig. 3.4(a)). To *fourth order*, w contains the frequencies $4\omega_1$, $4\omega_2$, $2\omega_1 + \omega_1 - \omega_1 = 2\omega_1$, $2\omega_2 + \omega_2 - \omega_2 = 2\omega_2$, $2\omega_1 - 2\omega_1$, $2\omega_2 - 2\omega_2$, $2\omega_1 - \omega_1 \pm \omega_2 = \omega_1 \pm \omega_2$, $3\omega_1 \pm \omega_2$, and $3\omega_2 \pm \omega_1$. The components $2\omega_1 - 2\omega_1$ and $2\omega_2 - 2\omega_2$ have a finite cycle-average but do again not lead to a dependence on the phase φ, since their phase is given by $2\varphi - 2\varphi = 0$. These terms are proportional to $\tilde{E}_1^4$ and $\tilde{E}_2^4$, respectively, hence they are proportional to the square of the light intensity of beam 1 and 2, respectively, and correspond to two-photon absorption[4]. *In contrast, the component* $3\omega_1 - \omega_2$ *oscillates with phase* $3\varphi - \varphi = 2\varphi$ *and has a finite cycle average if the condition* $\omega_2/\omega_1 = 3$ *is fulfilled* (in terms of bandwidth this would correspond to a subcycle pulse, which has not been achieved to date, also see Problem 2.2). This contribution to the fourth-order inversion is proportional to $\tilde{E}_1^3\tilde{E}_2$ and leads to an oscillation of the inversion versus φ with period π. It can be interpreted as an interference of one-photon (ω_2) and two-photon (ω_1) absorption. Similarly, for an *arbitrary even order* $N \geq 4$, the inversion w has a contribution at frequency $(\frac{N}{2}+1)\omega_1 - (\frac{N}{2}-1)\omega_2$, which becomes dc if

$$\frac{\omega_2}{\omega_1} = \frac{N+2}{N-2} \rightarrow 1 \quad \text{for } N \rightarrow \infty. \qquad (3.28)$$

---

[4] This type of two-photon absorption is distinct from that discussed in most textbooks. There, mostly transitions are considered that have a vanishing dipole matrix element d, i.e., where one-photon absorption is absent. There, E^2 drives the transition and, thus, two-photon resonance occurs at $\Omega = 2\omega_0$ rather than at $\Omega = 3\omega_0$ for the case discussed here.

Its phase is again $(\frac{N}{2}+1)\,\varphi - (\frac{N}{2}-1)\,\varphi = 2\varphi$. For example, for $N = 6$, this contribution occurs for $\omega_2/\omega_1 = 2$ (one octave). It can be interpreted as an interference of two- and three-photon absorption. From (3.28) we expect that the frequency components ω_1 and ω_2 have to be less separated in energy for large orders N, i.e., we anticipate that "shorter" ("longer") pulses require lower (higher) laser intensities to lead to a dependence on the phase φ. Also, the N-th–order contributions $(\frac{N}{2}+M)\,\omega_1 - (\frac{N}{2}-M)\,\omega_2$ with integer $M < N/2$ and phase $(\frac{N}{2}+M)\,\varphi - (\frac{N}{2}-M)\,\varphi = 2M\varphi$ exist, which lead to interference components of w versus φ with period π/M rather than π, provided the driving frequencies follow the relation $\omega_2/\omega_1 = \frac{N+2M}{N-2M}$. For example, $N = 6$ and $M = 2$ could be interpreted as an interference of one- and three-photon absorption.

It should be clear that the amplitudes of the various contributions depend strongly on the resonance conditions, i.e., on the ratios Ω/ω_1 and Ω/ω_2 (also see dependence on Ω/ω_0 in Figs. 3.9 and 3.10).

This reasoning has shown that, for few-cycle laser pulses, interference of different multiphoton pathways and a corresponding dependence of the inversion w after the pulse on the CEO phase ϕ only occurs at high orders N perturbation theory in the incident electric field. For a pulse spectrum covering one octave, we have $N \geq 6$. For the parameters of Figs. 3.9 and 3.10, the laser spectrum supports a maximum ratio of $\omega_2/\omega_1 \approx 1.65$. This means that the modulation occurs in $N \geq$ 10th-order perturbation theory according to (3.28) because $\frac{10+2}{10-2} = 1.5 < 1.65$. As the strength of these high orders must be comparable to that of the lower orders to make the interference contrast appreciable in magnitude, we conclude that the nonperturbative regime with $\Omega_R/\omega_0 \approx 1$ has to be reached to obtain sizable effects. This finding is consistent with the numerical calculations. Indeed, going from a peak Rabi frequency of $\Omega_R/\omega_0 = 1$ (Fig. 3.9) down to $\Omega_R/\omega_0 = 0.5$ (Fig. 3.10) reduces the relative modulation depth by a factor of a few hundred.

Box-shaped pulses

The carrier-envelope phase dependence of the inversion does, however, also strongly depend on the *shape* of the pulse envelope. Box-shaped envelopes containing an integer number of optical cycles (see, e.g., Fig. 3.4 where $\phi = 0$) are an exceptional and somewhat unrealistic but not unphysical case (the dc component is strictly zero), which actually turns out to be instructive. Let us consider a single-cycle pulse starting at $t = 0$, i.e., $\Omega_R(t) = \Omega_R \sin(\omega_0 t + \phi)$ for t in the interval $[0, 2\pi/\omega_0]$ and 0 else. For small Rabi frequency Ω_R, we can treat the light field as a perturbation and solve the Bloch equations (3.17) starting from the ground state up to *second order* in the incident field, equivalent to one-photon absorption. The mathematical procedure is as for the two-color excitation. Transverse as well as longitudinal damping are neglected. A straightforward calculation gives the upper-state occupation $f_2 = f_2(t \to \infty) = f_2(2\pi/\omega_0) = (w(2\pi/\omega_0) + 1)/2$ after the pulse

$$f_2 = \frac{4\,\Omega_R^2\,\Omega^2}{\left(\Omega^2 - \omega_0^2\right)^2}\,\sin^2\left(\pi\,\frac{\Omega}{\omega_0}\right)\left[1 + \frac{\omega_0^2 - \Omega^2}{\Omega^2}\,\sin^2\left(\pi\,\frac{\Omega}{\omega_0} + \phi\right)\right]. \quad (3.29)$$

Again, the periodicity of f_2 versus CEO phase ϕ is π. Note that the relative modulation depth does not depend on the Rabi frequency although f_2 itself is obviously proportional to the intensity of light. The modulation depth does depend on the transition frequency Ω and changes sign when Ω crosses the carrier frequency ω_0. For box-shaped pulses containing an integer number N of cycles of light, each π in (3.29) has to be replaced by $N\pi$. Amazingly, the CEO-phase dependence survives for arbitrary $N = 1, 2, 3, ...$ Thus, for box-shaped optical pulses, a CEO-phase dependence is expected even for a very weak ($\Omega_R/\omega_0 \to 0$) optical pulse containing many cycles of light ($N \to \infty$).

This seemingly surprising result is a consequence of the discontinuous jumps in the incident electric field $E(t)$ for all ϕ except for $\phi = 0, \pi, 2\pi, ...$ In the Fourier domain, broad spectral wings result and the maximum centered at $\omega = \omega_0$ interferes with that at $\omega = -\omega_0$. Consequently, even the laser spectrum $|E(\omega)|$ itself strongly depends on the CEO phase for these pulses (also, see Example 2.4). Hence, the CEO phase could simply be determined with a spectrometer. The two-level system just acts as a spectral filter that picks out a certain frequency range. Smoothing of the temporal jumps reduces the relative modulation depth and gradually brings one back towards the more realistic and more meaningful behavior for the pulse envelopes discussed above, where the laser spectrum does *not* depend on ϕ.

More details on the CEO-phase-dependent inversion can be found in Refs. [75–78].

3.6 High Harmonics from Two-Level Systems

We now want to approach yet larger Rabi frequencies, in which case the response tends to become rather complex. The simplest and cleanest situation is given when the envelope Rabi frequency is either constant or zero, as was the case in Fig. 3.5. Let us start our discussion with such box-shaped optical pulses (illustrated in Fig. 3.4(a)), which are $N = 30$ optical cycles in duration (e.g., for $\hbar\omega_0 = 1.5\,eV$, this would roughly correspond to a 90-fs long pulse). Damping is irrelevant under the conditions of this section unless the damping rate becomes comparable to the transition frequency, in which case no resonance exists at all. For our numerical calculations, we chose a transverse relaxation time T_2 corresponding to $\hbar\omega_0 T_2 = 1.5\,eV\,50\,fs$. Longitudinal damping is neglected ($T_1 = \infty$). To get an overview we can either fix Ω/ω_0 and depict the radiated intensity versus Ω_R/ω_0 and ω/ω_0 (Fig. 3.11) or, alternatively, fix Ω_R/ω_0 and plot the signals versus Ω/ω_0 and ω/ω_0 (Fig. 3.12).

For $\Omega_R/\omega_0 \ll 1$ on the vertical axis of Fig. 3.11(a), where $\Omega/\omega_0 = 1$ (resonant excitation), conventional Rabi flopping [53] occurs and the well-known Mollow triplet [56] can be seen at $\omega/\omega_0 = 1$ on the horizontal axis. At larger Ω_R/ω_0 approaching unity, carrier-wave Rabi flopping takes place and additional carrier-wave Mollow triplets appear around odd integers ω/ω_0. Beyond $\Omega_R/\omega_0 = 1$, the Mollow sidebands are "repelled" by the central peaks of the adjacent Mollow triplets. They oscillate around even integer values of ω/ω_0 and finally converge towards these values in the limit $\Omega_R/\omega_0 \gg 1$. On the way, they periodically cross at even integers ω/ω_0 with a

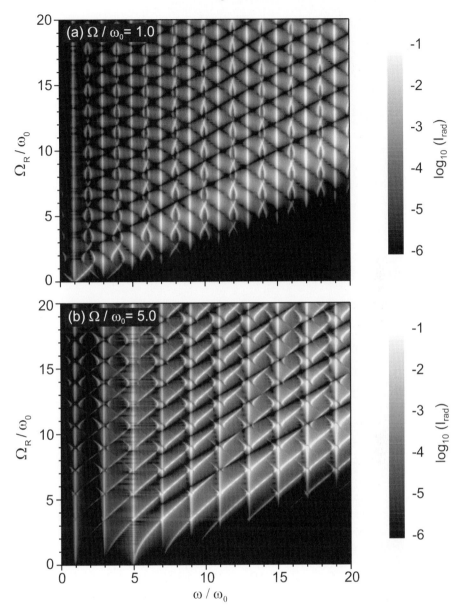

Fig. 3.11. Grayscale image of the radiated intensity $I_{rad} \propto |\omega^2 u(\omega)|^2$ from exact numerical solutions of the two-level system Bloch equations (3.17) versus peak Rabi frequency Ω_R and spectrometer frequency ω for two fixed values of the transition frequency Ω, all in units of the laser carrier frequency ω_0. The exciting box-shaped optical pulses according to Fig. 3.4(a) are $N = 30$ cycles long. **(a)** $\Omega/\omega_0 = 1$, **(b)** $\Omega/\omega_0 = 5$. Compare with Fig. 3.5. Reprinted with permission from T. Tritschler et al., Phys. Rev. A **68**, 033404 (2003) [79]. Copyright (2003) by the American Physical Society.

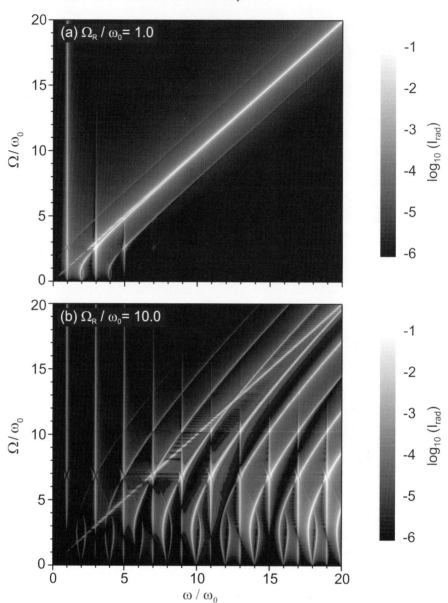

Fig. 3.12. As Fig. 3.11, but versus transition frequency Ω and spectrometer frequency ω for two fixed values of the peak Rabi frequency Ω_R, all in units of the laser carrier frequency ω_0. (a) $\Omega_R/\omega_0 = 1$, (b) $\Omega_R/\omega_0 = 10$. Compare with Figs. 3.7 and 3.6. Reprinted with permission from T. Tritschler et al., Phys. Rev. A **68**, 033404 (2003) [79]. Copyright (2003) by the American Physical Society.

period versus Ω_R/ω_0 of $\pi/2$ for $\Omega_R/\omega_0 \gg 1$ (whereas the first crossing occurs at $\Omega_R/\omega_0 \approx 1$. For off-resonant excitation, e.g. $\Omega/\omega_0 = 5$ in Fig. 3.11(b), the behavior is different for $\Omega_R/\omega_0 < 1$ and $\Omega_R/\omega_0 \approx 1$, but becomes similar to Fig. 3.11(a) for $\Omega_R/\omega_0 \gg 1$.

The other way to look at the parameter space is to fix the Rabi frequency Ω_R/ω_0. For large Ω/ω_0 but not too large peak Rabi frequencies Ω_R/ω_0 in Fig. 3.12, well-separated high harmonics are observed, as expected from our discussion on phenomenological nonlinear optics in Sect. 2.4. On the diagonal, where $\omega = \Omega$, very large resonant enhancement effects are observed. This is also true for the adjacent harmonics at spectrometer frequencies $\omega = \Omega \pm 2M\omega_0$ with integer M, which altogether leads to a band of enhancement around the diagonal in Fig. 3.12. Especially note that large contributions can occur at the spectral positions of even harmonics, as already discussed for Fig. 3.11. These contributions are especially pronounced for even integer values of Ω/ω_0 [67].

For more realistic smoothed box-shaped optical pulses the overall behavior remains the same. If, e.g., the electric-field envelope is switched on and off by Fermi functions rising and decaying within a few optical cycles, respectively, the behavior of Fig. 3.12 is slightly smeared out and the contributions at spectrometer frequencies $\omega = \Omega \pm 2M\omega_0$ decay more rapidly for large values of Ω/ω_0 as compared to box-shaped optical pulses.

For other pulse envelopes, the envelope Rabi frequency is not constant within the pulse, which effectively averages along the vertical axis of Fig. 3.11. This is further illustrated in Fig. 3.13 for the example of Gaussian pulses with an electric-field envelope given by $\tilde{E}(t) = \tilde{E}_0 \exp(-(t/t_0)^2)$. The temporal full width at half-maximum (FWHM) of the intensity profile is given by $t_{\text{FWHM}} = t_0\, 2\sqrt{\ln\sqrt{2}}$ and translates into a FWHM of $N = t_{\text{FWHM}}/(2\pi/\omega_0)$ optical cycles in Fig. 3.13. In (a), where $N = 30$ and $\phi = 0$, the anticipated averaging can be seen clearly. As a result, the contributions at odd integers ω/ω_0 have almost disappeared in favor of even contributionss. This is exactly opposite to what one might have expected intuitively. For $N = 30$ and $\phi = \pi/2$ (not shown), the behavior is similar – apart from fine details, which are hardly visible on this scale. For few-cycle optical pulses (Fig. 3.13(b)), the various contributions get largely broadened spectrally and their mutual interference leads to rather "messy" spectra, which have lost all of the beautiful fine details of Fig. 3.11. It is clear that this interference also introduces a dependence on the carrier-envelope phase ϕ as discussed in more detail in Sects. 7.1 and 7.2.

One might be tempted to argue that the peaks at even integers ω/ω_0 in the optical spectra at large Ω_R/ω_0 arise from the fact that the complete system, i.e., two-level system plus light field, no longer has inversion symmetry at large electric fields. This reasoning is, however, not consistent with the Bloch equations. Space inversion means that we have to replace $r \to -r$. As a result, the dipole matrix element transforms as $d \to -d$ and the electric field as $E(t) \to -E(t)$. Hence, we have $\hbar^{-1}dE(t) = \Omega_R(t) \to +\Omega_R(t)$ and the Bloch equations (3.17) remain completely unchanged under space inversion. Thus, the solution for the Bloch vector $(u(t), v(t), w(t))^{\text{T}}$ is also unchanged. Finally, the macroscopic optical polarization, which is given by

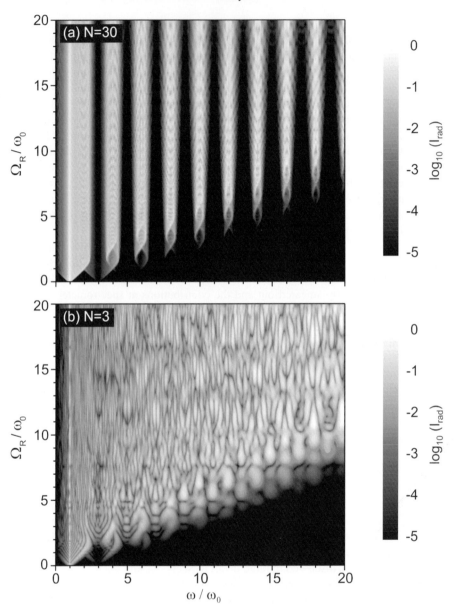

Fig. 3.13. As Fig. 3.11(a), but for Gaussian optical pulses with CEO phase $\phi = 0$ and with a full width at half-maximum of $t_{\text{FWHM}} = N\ 2\pi/\omega_0$. **(a)** $N = 30$, **(b)** $N = 3$. Reprinted with permission from T. Tritschler et al., Phys. Rev. A **68**, 033404 (2003) [79]. Copyright (2003) by the American Physical Society.

$P(t) = N_{2LS}/V\,d\,u(t)$ transforms according to $P(t) \to -P(t)$. Consequently, in an expansion of the polarization in terms of powers of the electric field up to infinite order, strictly no even orders occur – even for arbitrarily large electric fields. Thus, we deliberately avoid to call a peak at, e.g., $\omega/\omega_0 = 2$, second harmonic generation (SHG). As discussed in Sect. 2.5, a strict definition of SHG must be based on its carrier frequency or its phase, which are $2\omega_0$ and 2ϕ, respectively. The peaks at $\omega/\omega_0 = 2$ from the two-level system are not consistent with this definition. Thus, we rather use the notion "third-harmonic generation in disguise of second-harmonic generation" for this unusual contribution. A similar argument obviously holds for all the other even integers ω/ω_0, which must not be called *even harmonics* in the framework of nonlinear optics. In some of the literature – written before the importance of the carrier-envelope phase ϕ was fully appreciated – these peaks are nevertheless called *even harmonics*.

Exact numerical solutions of the two-level system Bloch equations (3.17) – as discussed in this chapter so far – are actually very simple and quick with the computers and software packages at hand today. Thus, there is no need for simplified numerical treatments. The spirit of the following approximate schemes rather is to give additional insight into the physics or to deliver "handy" analytical expressions.

3.6.1 The "Static-Field Approximation"

Is there anything that can easily be evaluated on a piece of paper rather than on a computer? For times much shorter than a cycle of light, $2\pi/\omega_0$, we can employ the "static-field approximation", i.e., we approximate $\Omega_R(t) = \Omega_R$ as constant in time. In this limit, the "optical" transitions can obviously be thought of as originating from electrostatic tunneling of electrons from the ground state into the excited states and vice versa. We will see below, for which field amplitudes or Rabi frequencies this limit becomes meaningful. Within this limit, it is straightforward to solve the Bloch equations (3.17) analytically [22] (also see Problem 3.7). This leads to the Bloch vector

$$\begin{pmatrix} u(t) \\ v(t) \\ w(t) \end{pmatrix} = \mathcal{M}(t) \begin{pmatrix} u(0) \\ v(0) \\ w(0) \end{pmatrix}, \tag{3.30}$$

with the (3×3) rotation matrix

$$\mathcal{M}(t) = \begin{pmatrix} \dfrac{4\Omega_R^2 + \Omega^2\cos(\Omega_{\mathrm{eff}}t)}{\Omega_{\mathrm{eff}}^2} & \dfrac{\Omega}{\Omega_{\mathrm{eff}}}\sin(\Omega_{\mathrm{eff}}t) & \dfrac{2\Omega\Omega_R}{\Omega_{\mathrm{eff}}^2}(\cos(\Omega_{\mathrm{eff}}t)-1) \\[3ex] -\dfrac{\Omega}{\Omega_{\mathrm{eff}}}\sin(\Omega_{\mathrm{eff}}t) & \cos(\Omega_{\mathrm{eff}}t) & -\dfrac{2\Omega_R}{\Omega_{\mathrm{eff}}}\sin(\Omega_{\mathrm{eff}}t) \\[3ex] \dfrac{2\Omega\Omega_R}{\Omega_{\mathrm{eff}}^2}(\cos(\Omega_{\mathrm{eff}}t)-1) & \dfrac{2\Omega_R}{\Omega_{\mathrm{eff}}}\sin(\Omega_{\mathrm{eff}}t) & \dfrac{\Omega^2 + 4\Omega_R^2\cos(\Omega_{\mathrm{eff}}t)}{\Omega_{\mathrm{eff}}^2} \end{pmatrix}. \tag{3.31}$$

Obviously, the optical polarization $P(t) \propto u(t)$ as well as the other two components of the Bloch vector oscillate with the effective frequency Ω_{eff}, which is given by

$$\Omega_{\text{eff}} = \sqrt{4\Omega_R^2 + \Omega^2} \,. \tag{3.32}$$

Remember that this *"static-field approximation"* is only justified for times $t \ll 2\pi/\omega_0$, hence it is relevant in the limit $\Omega_{\text{eff}} \gg \omega_0$. It can be viewed as the opposite of the rotating-wave approximation (see Sect. 3.3). There, almost nothing is supposed to happen on the timescale of light, whereas here all the significant dynamics takes place within an optical cycle. For $\Omega_R \gg \Omega$, we have $\Omega_{\text{eff}} = 2\Omega_R$, which means that twice the peak Rabi frequency is the largest occurring frequency, hence, the highest harmonic generated, *the cutoff*, is given by

$$N_{\text{cutoff}} = 2\,\frac{\Omega_R}{\omega_0} \tag{3.33}$$

(compare black areas on the lower RHS in Figs. 3.11 (a) and (b)).

Starting from the ground state, i.e., from Bloch vector $(0, 0, -1)^{\text{T}}$ at time $t = 0$, the inversion according to (3.30) and (3.31) is given by

$$w(t) = -\frac{\Omega^2 + 4\Omega_R^2 \cos(\Omega_{\text{eff}} t)}{\Omega_{\text{eff}}^2}\,. \tag{3.34}$$

Thus, the two-level system can even perform Rabi flopping for far off-resonant conditions, i.e. for $\Omega \gg \omega_0$, if the intensity is so large that it roughly corresponds to a Rabi frequency of $\Omega_R = \Omega$. With (3.32), this leads to $\Omega_{\text{eff}} = \sqrt{5}\,\Omega_R$, hence

$$w(t) = -\frac{1}{5} - \frac{4}{5}\cos\left(\sqrt{5}\,\Omega_R t\right)\,. \tag{3.35}$$

The maximum inversion is $w = +3/5$, corresponding to $f_2 = 80\%$ occupation of the excited state. In a quantum optical description of the light field, this behavior can be interpreted as carrier-wave Rabi flopping due to *multiphoton absorption*. For yet larger intensities, i.e., in the limit $\Omega_R \gg \Omega$, (3.34) simplifies to

$$w(t) = -\cos(2\Omega_R t)\,, \tag{3.36}$$

and even 100% maximum inversion results. Equation (3.36) obviously resembles Rabi's famous result obtained within the rotating-wave approximation, i.e., (3.21) with the envelope Rabi frequency $\tilde{\Omega}_R$ replaced by $2\Omega_R$.

Rabi flopping within the static-field limit will soon reappear when discussing electron–positron pair generation in vacuum in Sect. 4.5. Moreover, we will re-encounter the static-field or high-intensity limit several times below, e.g., in the context of the Keldysh parameter and field-ionization of atoms in Sects. 5.2 and 5.3 or for the dynamic Franz–Keldysh effect in Sect. 7.3.

Problem 3.7. Show that (3.30) together with (3.31) and (3.32) holds under these conditions. To get there, first express the Bloch equations (3.17) for $\Omega_R(t) = \Omega_R = $ const. as two coupled harmonic oscillator equations for the optical polarization u and the inversion w.

3.6.2 The "Square-Wave Approximation"

The Bloch equations (3.17) describe rotations of the Bloch vector on the Bloch sphere. Within the regime of extreme nonlinear optics, the behavior becomes "enriched" by the fact that one of the rotation frequencies, namely $2\Omega_R(t)$, itself oscillates with the carrier frequency of light and periodically changes sign. This oscillation is sinusoidal, yet, one might ask whether it is really so important that it is sinusoidal. Having in mind what we have said about the "static-field approximation" in Sect. 3.6.1, it is straightforward to extend this result to piecewise constant electric fields $E(t)$ or Rabi frequencies $\Omega_R(t)$, respectively [79]. This leads us to investigating the "square-wave approximation" in which we approximate the Rabi frequency for a constant envelope via

$$\Omega_R(t) = \Omega_R \cos(\omega_0 t + \phi) \rightarrow \frac{2}{\pi} \Omega_R \, \mathrm{sgn}(\cos(\omega_0 t + \phi)), \tag{3.37}$$

where the signum function is defined as $\mathrm{sgn}(x) = +1$ for $x > 0$, $\mathrm{sgn}(x) = -1$ for $x < 0$ and $\mathrm{sgn}(x) = 0$ for $x = 0$. The prefactor $2/\pi$ ensures that the average Rabi frequency within half an optical cycle is the same for the "square-wave approximation" and the exact problem. In that half of the optical cycle where the Rabi frequency is positive (negative), the Bloch vector rotates via the matrix $\mathcal{M}_+$ ($\mathcal{M}_-$), where $\mathcal{M}_\pm$ results from $\mathcal{M}$ by replacing $\Omega_R \rightarrow \pm(2/\pi)\,\Omega_R$ in (3.31) and (3.32). For more than half an optical cycle, the dynamics of the Bloch vector is described by

$$\begin{pmatrix} u(t) \\ v(t) \\ w(t) \end{pmatrix} = \mathcal{M}_{\mathrm{tot}}(t) \begin{pmatrix} u(0) \\ v(0) \\ w(0) \end{pmatrix}, \tag{3.38}$$

where the total matrix $\mathcal{M}_{\mathrm{tot}}$ is a product of simple analytical (3×3) rotation matrices: For times t after the optical pulse with integer number of cycles of light N, where $\Omega_R(t) = 0$, we have

$$\mathcal{M}_{\mathrm{tot}}(t) = \mathcal{M}_0\left(t - N\frac{2\pi}{\omega_0}\right)\left(\mathcal{M}_-\left(\frac{\pi}{\omega_0}\right)\mathcal{M}_+\left(\frac{\pi}{\omega_0}\right)\right)^N, \tag{3.39}$$

where $\mathcal{M}_0$ results from $\mathcal{M}$ by replacing $\Omega_R \rightarrow 0$ in (3.31) and (3.32). $\mathcal{M}_0$ describes only a rotation in the uv-plane with the transition frequency Ω and can be simplified to

$$\mathcal{M}_0(t) = \begin{pmatrix} \cos(\Omega t) & +\sin(\Omega t) & 0 \\ -\sin(\Omega t) & \cos(\Omega t) & 0 \\ 0 & 0 & 1 \end{pmatrix}. \tag{3.40}$$

Within the optical pulse, we obtain for times t with $\Omega_R(t) > 0$

$$\mathcal{M}_{\text{tot}}(t) = \mathcal{M}_+\left(t - N_t \frac{2\pi}{\omega_0}\right)\left(\mathcal{M}_-\left(\frac{\pi}{\omega_0}\right)\mathcal{M}_+\left(\frac{\pi}{\omega_0}\right)\right)^{N_t}, \tag{3.41}$$

and for times t with $\Omega_R(t) < 0$

$$\mathcal{M}_{\text{tot}}(t) = \mathcal{M}_-\left(t - \left[N_t + \frac{1}{2}\right]\frac{2\pi}{\omega_0}\right)\mathcal{M}_+\left(\frac{\pi}{\omega_0}\right)\left(\mathcal{M}_-\left(\frac{\pi}{\omega_0}\right)\mathcal{M}_+\left(\frac{\pi}{\omega_0}\right)\right)^{N_t}. \tag{3.42}$$

Here we have introduced the integer number of cycles N_t completed up to time t, which is given by

$$N_t = \text{int}\left(\frac{\omega_0 t}{2\pi}\right). \tag{3.43}$$

The value of the integer function $\text{int}(x)$ is given by the largest integer $\leq x$.

We first test the "square-wave approximation" by depicting its solutions in Fig. 3.14. Parameters are identical to those of the exact numerical calculations in Fig. 3.11, which allows for a direct comparison. The overall qualitative agreement is amazing, especially for the (a) parts. There, $\Omega/\omega_0 = 1$ (resonant excitation), which is just the generalization of Rabi flopping and Mollow triplets. For instance, the periodically occurring constrictions of the repelling Mollow sidebands at even integers ω/ω_0 versus Ω_R/ω_0 with period $\pi/2$ (see discussion in Sect. 3.6) are very nicely reproduced. For off-resonant excitation ($\Omega/\omega_0 = 5$) in (b), the square-wave approximation is less convincing. This aspect can be understood intuitively. For resonant excitation ($\Omega/\omega_0 = 1$), the transition frequency resonantly enhances those frequency components of the square-wave with frequency ω_0. Thus, the artificial higher harmonics of the square-wave at frequencies $3\omega_0$, $5\omega_0$, ... are relatively suppressed. Clearly, the "square-wave approximation" does not properly recover the limit of linear optics, in the sense that $u(t)$ is not sinusoidal in that limit (as it should be), equivalent to higher harmonics of the carrier frequency ω_0 in the Fourier domain. Thus, the lower RHS of Figs. 3.14(a) and (b) (which is dark in Figs. 3.11(a) and (b)) is an obvious artifact of the "square-wave approximation". This artifact is unimportant because we are rather interested in the regime of extreme nonlinear optics.

The simplest cases of commensurability of the frequencies ω_0, Ω, and Ω_R within the "square-wave approximation" are given by

$$\Omega_{\text{eff}} \frac{\pi}{\omega_0} = M\, 2\pi, \tag{3.44}$$

with integer M, for which we have

$$\mathcal{M}_+\left(\frac{\pi}{\omega_0}\right) = \mathcal{M}_-\left(\frac{\pi}{\omega_0}\right) = \begin{pmatrix} 1 & 0 & 0 \\ 0 & 1 & 0 \\ 0 & 0 & 1 \end{pmatrix}. \tag{3.45}$$

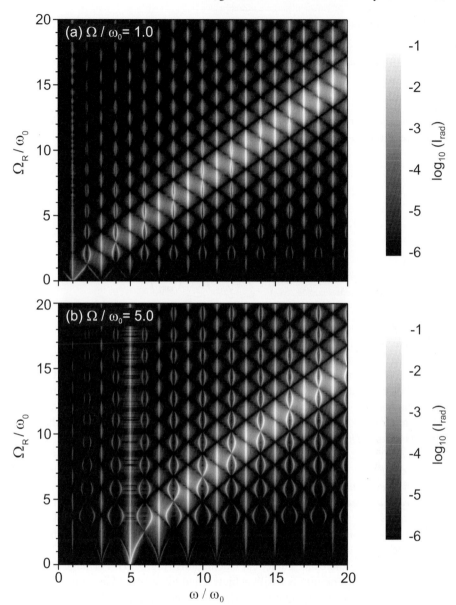

Fig. 3.14. As Fig. 3.11, but based on the analytical solution of the two-level system Bloch equations within the "square-wave approximation". (a) $\Omega/\omega_0 = 1$, (b) $\Omega/\omega_0 = 5$. Reprinted with permission from T. Tritschler et al., Phys. Rev. A **68**, 033404 (2003) [79]. Copyright (2003) by the American Physical Society.

Under these conditions, an integer number of Rabi flops is completed after half an optical cycle π/ω_0. Inserting

$$\Omega_{\text{eff}} = \sqrt{4\left(\frac{2}{\pi}\,\Omega_R\right)^2 + \Omega^2} \qquad (3.46)$$

into (3.44), we get that *commensurability occurs for specific Rabi frequencies* given by the relation

$$\frac{\Omega_R}{\omega_0} = \frac{\pi}{2}\sqrt{M^2 - \frac{1}{4}\left(\frac{\Omega}{\omega_0}\right)^2}, \qquad (3.47)$$

with $M = 1, 2, 3, \ldots$ For these Rabi frequencies, *peaks at even integers*

$$\frac{\omega}{\omega_0} = \frac{\Omega_{\text{eff}}}{\omega_0} = \sqrt{\frac{16}{\pi^2}\left(\frac{\Omega_R}{\omega_0}\right)^2 + \left(\frac{\Omega}{\omega_0}\right)^2} = 2\,M \qquad (3.48)$$

are observed in the optical spectrum – apart from the less interesting peaks at odd integers ω/ω_0, which also occur in traditional nonlinear optics. These peaks at even integers ω/ω_0 form the bright band in Fig. 3.14, whereas the other even integers ω/ω_0 are absent in the spectrum. This bright band also occurs in the exact numerical solutions (Fig. 3.11). There, in contrast to the "square-wave approximation", the instantaneous Rabi frequency $\Omega_R(t)$ varies within half an optical cycle (somewhat similar to a "chirped" optical pulse), which also introduces peaks at other even integers ω/ω_0. *This shows that the constrictions formed by the crossing Mollow triplets in Fig. 3.11 can be interpreted as points of commensurability of the carrier frequency of light ω_0, the transition frequency Ω, and the peak Rabi frequency Ω_R. Here an integer number of Rabi flops is completed after half an optical cycle and, thus, peaks at even integers ω/ω_0 occur in the optical spectrum.* For example, for $M = 1$ and $\Omega/\omega_0 = 1$ in (3.47), we get $\Omega_R/\omega_0 = \sqrt{3}\,\pi/4 \approx 1.36$ (Fig. 3.14), which roughly agrees with $\Omega_R/\omega_0 \approx 1$ in the exact numerical calculations (Fig. 3.11). For integers $M \gg \Omega/\omega_0$ we get $\Omega_R/\omega_0 = M\,\pi/2$. This period of $\pi/2$ is also precisely found in the exact numerical calculations (Fig. 3.11). For large Rabi frequencies, commensurability is easily achieved and these "even harmonics" become the rule rather than the exception – despite the fact that the two-level system has inversion symmetry. In between these points of commensurability, it takes some optical cycles to again approach the initial state. In the Fourier domain, this obviously corresponds to nearby sidebands around those even integers ω/ω_0 (see Figs. 3.11 and 3.14).

3.6.3 The Dressed Two-Level System: Floquet States

Floquet's theorem [80] generally allows us to translate any time-dependent Hamiltonian that is strictly periodic in time to an effective *time-independent* Hamiltonian represented by an infinite matrix [81]. Starting from this point, all known approximative schemes for stationary Hamiltonians can be applied.

Let us derive the corresponding time-independent matrix for the two-level system. First, we make the obvious ansatz

$$\psi(t) = a_1(t)\,\psi_1 + a_2(t)\,\psi_2\,, \tag{3.49}$$

where ψ_1 and ψ_2 are the ground and excited state of the two-level system, respectively (see Sect. 3.2). With this ansatz, the Schrödinger equation $i\hbar\dot{\psi}(t) = \mathcal{H}\psi(t)$ can be expressed in (2×2) matrix form as

$$i\hbar\frac{\partial}{\partial t}\begin{pmatrix} a_1 \\ a_2 \end{pmatrix} = \begin{pmatrix} E_1 & -\hbar\Omega_R(t) \\ -\hbar\Omega_R(t) & E_2 \end{pmatrix}\begin{pmatrix} a_1 \\ a_2 \end{pmatrix} = \mathcal{H}\begin{pmatrix} a_1 \\ a_2 \end{pmatrix}\,, \tag{3.50}$$

where the time-dependent Rabi frequency (see (3.18) and (3.19)) is again given by

$$\Omega_R(t) = \Omega_R\cos(\omega_0 t + \phi) = \frac{\Omega_R}{2}\left(e^{+i(\omega_0 t+\phi)} + e^{-i(\omega_0 t+\phi)}\right)\,. \tag{3.51}$$

For continuous-wave excitation, Ω_R is constant in time. Along the lines of our discussion in Sect. 3.6, we expect that the coefficients $a_n(t)$ comprise harmonics of the laser carrier frequency ω_0. This leads us to the ansatz

$$a_n(t) = e^{-i\omega t}\sum_{N=-\infty}^{+\infty} a_{n,N}\,e^{-iN(\omega_0 t+\phi)} \tag{3.52}$$

for the two levels $n = 1, 2$. Introducing this ansatz into (3.50) and ordering the terms according to their exponentials immediately delivers an infinite set of coupled linear equations for the *time-independent* coefficients $a_{n,N}$. Ordering them into a vector according to

$$\psi' = (..., a_{1,-1}, a_{2,-1}, a_{1,0}, a_{2,0}, a_{1,1}, a_{2,1}, a_{1,2}, a_{2,2}, ...)^\mathrm{T}\,, \tag{3.53}$$

inserting the transition energy $\hbar\Omega = (E_2 - E_1)$ from (3.11) and choosing $E_1 = 0$ without loss of generality, leads to the eigenvalue problem

$$\mathcal{H}'\psi' = \hbar\omega\,\psi'\,, \tag{3.54}$$

with the *stationary*, infinite Floquet matrix

$$\mathcal{H}' = \hbar \begin{pmatrix}
\cdots & \cdots & \cdots & \cdots & \cdots & \cdots & \cdots & \cdots & \cdots & \cdots \\
\cdots & +\omega_0 & 0 & 0 & -\dfrac{\Omega_R}{2} & 0 & 0 & 0 & 0 & \cdots \\
\cdots & 0 & \Omega+\omega_0 & -\dfrac{\Omega_R}{2} & 0 & 0 & 0 & 0 & 0 & \cdots \\
\cdots & 0 & -\dfrac{\Omega_R}{2} & \mathbf{0} & 0 & 0 & -\dfrac{\Omega_R}{2} & 0 & 0 & \cdots \\
\cdots & -\dfrac{\Omega_R}{2} & 0 & 0 & \Omega & -\dfrac{\Omega_R}{2} & 0 & 0 & 0 & \cdots \\
\cdots & 0 & 0 & 0 & -\dfrac{\Omega_R}{2} & -\omega_0 & 0 & 0 & -\dfrac{\Omega_R}{2} & \cdots \\
\cdots & 0 & 0 & -\dfrac{\Omega_R}{2} & 0 & 0 & \Omega-\omega_0 & -\dfrac{\Omega_R}{2} & 0 & \cdots \\
\cdots & 0 & 0 & 0 & 0 & 0 & -\dfrac{\Omega_R}{2} & -2\omega_0 & 0 & \cdots \\
\cdots & 0 & 0 & 0 & 0 & -\dfrac{\Omega_R}{2} & 0 & 0 & \Omega-2\omega_0 & \cdots \\
\cdots & \cdots & \cdots & \cdots & \cdots & \cdots & \cdots & \cdots & \cdots & \cdots
\end{pmatrix}.$$

$$(3.55)$$

The zero in the fourth row and fourth column corresponding to $n = 1$ and $N = 0$ is highlighted. For small Rabi frequencies Ω_R, i.e., in the perturbative regime, the eigenfrequencies ω are approximately given by the diagonal elements of this matrix, namely by the frequencies 0, $\pm\omega_0$, $\pm2\omega_0$, $\pm3\omega_0$, ..., Ω, $\Omega \pm \omega_0$, $\Omega \pm 2\omega_0$, etc. We have already encountered them in Fig. 3.12(a). For larger Rabi frequency, the off-diagonal elements lead to couplings, hence the eigenfrequencies are modified. This especially leads to a variety of avoided crossings [55,82], which we have also already seen in Figs. 3.11 and 3.12(b) and that are illustrated schematically in a different way in Fig. 3.15. The eigenvectors of this stationary matrix (3.55) are often referred to as *Floquet states* – the eigenstates of the two-level system "dressed" by the light. Sometimes an expansion in terms of these Floquet states can be useful.

Within a quantum optical description of the light field in terms of a single quantized mode, the analogue of our classical treatment is the Jaynes–Cummings model [8, 83]. There, without two-level-system–photon interaction, e.g., the state with energy $\hbar\Omega + 2\hbar\omega_0$ can be interpreted as "one electron in the excited state plus two photons". With interaction, one again gets mixed states.

Problem 3.8. Draw the level diagram as in Fig. 3.15 for the case $\Omega = 3\,\omega_0$.

Problem 3.9. Arrange the two-level system Schrödinger equation (3.50) into the form of the known *Riccati nonlinear differential equation*, i.e., into the form

$$\frac{d\mathcal{R}}{dt} = \mathcal{A}_0(t) + \mathcal{A}_1(t)\,\mathcal{R} + \mathcal{A}_2(t)\,\mathcal{R}^2 , \qquad (3.56)$$

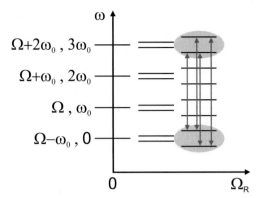

Fig. 3.15. Illustration of the eigenfrequencies of the Hamilton matrix (3.55) for resonant excitation, i.e., for $\Omega = \omega_0$. For example, the four optical transitions from the lowest energy doublet towards the highest energy doublet on the RHS (see arrows and gray ellipses) – two of which are degenerate – lead to the third-harmonic Mollow triplet, which we have already discussed in Sect. 3.3 and shown in Figs. 3.5 and 3.11(a).

with the complex function $\mathcal{R}(t)$ and the coefficients $\mathcal{A}_0(t)$, $\mathcal{A}_1(t)$, and $\mathcal{A}_2(t)$. In this form, the equation of motion can still not be solved analytically in general, but the Riccati equation has quite some interesting mathematical properties that can be taken advantage of.

4

The Drude Free-Electron Model and Beyond ...

4.1 Linear Optics: The Drude Model

A free electron with mass m_e and charge $-e$, driven by the laser electric field $E(t)$ obeys Newton's second law according to

$$m_e \ddot{x}(t) = -e\, E(t)\,, \tag{4.1}$$

where the coordinate x is again the electron displacement with respect to some fixed positive charge. This is simply (3.1) without a restoring force, i.e., with $\mathcal{D} = 0$ (compare Figs. 3.1 und 4.1). Thus, the *linear optical susceptibility* follows from (3.3) directly by setting $\mathcal{D} = \Omega = 0$. With the number of electrons N_e (replacing N_{osc}) per volume V we have

$$\chi(\omega) = -\frac{e^2 N_e}{\epsilon_0 V\, m_e}\frac{1}{\omega^2} = -\frac{\omega_{pl}^2}{\omega^2}\,. \tag{4.2}$$

Here we have introduced the *plasma frequency* ω_{pl} given by

$$\omega_{pl} = \sqrt{\frac{e^2 N_e}{\epsilon_0 V\, m_e}}\,. \tag{4.3}$$

Note that the plasma frequency increases with the square root of the electron density N_e/V. In a dielectric material, ϵ_0 has to be replaced by $\epsilon_0 \rightarrow \epsilon\epsilon_0$, with the relative dielectric constant ϵ.

If we have just an individual charge rather than a homogeneous gas of electrons, the oscillatory acceleration of the charge leads to Thomson scattering. It is analogous to Rayleigh scattering on an electric dipole – which makes the sky blue. Note that the frequency of the scattered light is identical to that of the incoming light, *linear Thomson scattering* is elastic scattering. The well-known radiation pattern of linear Thomson scattering is shown in Fig. 4.5(a).

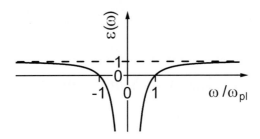

Fig. 4.1. Scheme of the linear dielectric function $\epsilon(\omega) = 1 + \chi(\omega)$ versus spectrometer frequency ω according to the Drude model, (4.2). ω_{pl} is the plasma frequency [268].

Can we associate a relevant energy to the intensity of the light field driving the electron? If yes, this energy would be the analogue of the Rabi energy for the two-level system. Let us consider a laser electric field $E(t)$ with constant envelope $\tilde{E}_0$, polarized along the x-direction, i.e.,

$$E(t) = \tilde{E}_0 \cos(\omega_0 t + \phi). \tag{4.4}$$

Inserting this expression into Newton's law (4.1) together with the initial condition $v(0) = 0$ delivers the electron velocity $v(t) = \dot{x}(t)$ according to

$$v(t) = -\frac{e\tilde{E}_0}{m_e \omega_0} \sin(\omega_0 t + \phi). \tag{4.5}$$

Averaging the classical electron kinetic energy $E_{kin}(t) = \frac{m_e}{2} v^2(t)$ over an optical cycle, $2\pi/\omega_0$, and remembering that $\langle \sin^2(\omega_0 t + \phi) \rangle = 1/2$, we obtain the *ponderomotive energy* (also known as wiggle or quiver energy)

$$\langle E_{kin} \rangle = \frac{1}{4} \frac{e^2 \tilde{E}_0^2}{m_e \omega_0^2}. \tag{4.6}$$

Obviously, the ponderomotive energy is directly proportional to the light intensity I ($\propto \tilde{E}_0^2$, see (2.16)). The peak kinetic energy is twice the ponderomotive energy.

▶ **Example 4.1.** For $\tilde{E}_0 = 4 \times 10^9$ V/m and GaAs parameters ($m_e = 0.07 \times m_0$, with the free electron mass $m_0 = 9.1091 \times 10^{-31}$ kg and $\hbar\omega_0 = E_g^{GaAs} = 1.42\,eV$) one obtains $\langle E_{kin} \rangle = 2.16\,eV$. For $\tilde{E}_0 = 6 \times 10^9$ V/m and ZnO parameters ($m_e = 0.24 \times m_0$ and $\hbar\omega_0 = 1.5\,eV$) we get $\langle E_{kin} \rangle = 1.27\,eV$.
By the way: For $\tilde{E}_0 = 4 \times 10^9$ V/m and GaAs parameters, the peak acceleration of the crystal electron, a_e^0, is given by $|a_e^0| = e/m_e\,\tilde{E}_0 = 1.0 \times 10^{22}$ m/s^2 $= 10^{21} \times$ g with the gravitational acceleration constant near the earth's surface g $= 9.81$ m/s^2. Compared with this, the maximum acceleration of a *Formula–1* race car, which is on the order of $10^1 \times$ g, is really negligible.   ◀

Problem 4.1. What are the peak (classical) charge displacements, x_0, for an electric field of $\tilde{E}_0 = 4 \times 10^9$ V/m for (a) Rabi oscillations and (b) free motion of an electron in a typical semiconductor ? How does x_0 scale with $\tilde{E}_0$ for the two cases ?

4.2 Electron Wave Packets Driven by Light

Quantum mechanically, the problem of an electron in a laser electric field (oscillating in time) is analogous to that of the light field with carrier frequency ω_0 in a mode-locked laser oscillator (see Sect. 2.3). There, the electromagnetic wave packet (the laser pulse) periodically oscillates back and forth between the laser mirrors (with the roundtrip frequency f_r). This leads to sidebands of ω_0 – the frequency comb. These sidebands are rigidly upshifted by the carrier-envelope offset frequency f_ϕ as a result of the phase slip $\Delta\phi$ of the electromagnetic wave packet from one roundtrip to the next according to (2.33).

4.2.1 Semiclassical Considerations

In analogy to this, semiclassically speaking, the electron wave packet in a periodic laser field acquires a quantum phase in one optical cycle, $\Delta\phi_e$, which is given by the cycle-average[1]

$$\Delta\phi_e = \langle 2\pi \, \frac{\left(v_{\text{phase}} - v_{\text{group}}\right) \dfrac{2\pi}{\omega_0}}{\lambda_e} \rangle \,, \tag{4.7}$$

with the electron de Broglie wavelength $\lambda_e = 2\pi/k_x$ and the period of light $2\pi/\omega_0$. With the dispersion relation of, e.g., vacuum electrons or crystal electrons in a solid within the effective-mass approximation

$$E_e(k_x) = \hbar\omega_e(k_x) = \frac{\hbar^2 k_x^2}{2\,m_e} \,, \tag{4.8}$$

and with $v_{\text{phase}} = \omega_e/k_x$ and $v_{\text{group}} = d\omega_e/dk_x$, we obtain the phase slip of the oscillating electron wave packet from one optical cycle to the next

$$\Delta\phi_e = \langle 2\pi \, \frac{\left(\dfrac{\hbar k_x}{2\,m_e} - 2\,\dfrac{\hbar k_x}{2\,m_e}\right)\dfrac{2\pi}{\omega_0}}{2\pi/k_x} \rangle$$

$$= -2\pi \, \frac{\langle \dfrac{\hbar^2 k_x^2(t)}{2\,m_e} \rangle}{\hbar\omega_0}$$

$$= -2\pi \, \frac{\langle E_{\text{kin}} \rangle}{\hbar\omega_0} \,. \tag{4.9}$$

---

[1] Note that λ_e, v_{phase}, and v_{group} vary in time via $k_x = k_x(t)$.

Note that the minus sign is due to the fact that the electron group velocity is larger than its phase velocity, while for photons the situation is usually reversed, i.e., their group velocity is smaller than their phase velocity.

According to (4.9), the phase slip becomes appreciable in magnitude if the ponderomotive energy approaches the carrier photon energy, i.e., if

$$\frac{\langle E_{\text{kin}} \rangle}{\hbar \omega_0} \approx 1 . \tag{4.10}$$

Note that this ratio scales as $1/\omega_0^3$. Furthermore, in analogy to the light field in a laser cavity, we expect that the density of states of the combined system *electron and light field* exhibits photon sidebands of the electron density of states at $\pm N \hbar \omega_0$ (with integer N), which – in analogy to (2.33) – are upshifted in energy [86–88] according to

$$\hbar \omega_0 \frac{|\Delta \phi_e|}{2\pi} = \langle E_{\text{kin}} \rangle , \tag{4.11}$$

i.e., upshifted by $\langle E_{\text{kin}} \rangle$. *Thus, the ponderomotive energy $\langle E_{\text{kin}} \rangle$ for electrons is analogous to the carrier-envelope offset frequency f_ϕ for photons.*

4.2.2 Quantum-Mechanical Treatment: Dressed Electrons

This reasoning is nice, simple and intuitive – but ignores one aspect of the problem, i.e., that the electron group velocity depends on frequency: We have a finite group velocity dispersion and anticipate a temporal broadening of any spatially localized wave packet. In order to address this aspect we have to solve the one-dimensional time-dependent (nonrelativistic) Schrödinger equation for the electron with charge $-e$, given by

$$i\hbar \frac{\partial}{\partial t} \psi(x, t) = \frac{1}{2m_e} (p_x + e A_x)^2 \, \psi(x, t) + V(x, t) \, \psi(x, t) , \tag{4.12}$$

with the momentum operator

$$p_x = -i\hbar \frac{\partial}{\partial x} . \tag{4.13}$$

From basic electrodynamics we know that the laser electric field $E(x, t)$ is related to the vector potential $A(x, t)$ and the electrostatic potential $\phi(x, t)$ (with $V(x, t) = -e\phi(x, t)$) via

$$E(x, t) = -\frac{\partial A(x, t)}{\partial t} - \frac{\partial \phi(x, t)}{\partial x} . \tag{4.14}$$

This gives us two attractive options: (i) *the radiation gauge*, i.e., $\phi(x, t) = 0$, and (ii) *the electric-field gauge*, i.e., set $A_x = 0$. We discuss both in the following. We will see later that – depending on the conditions and on the electric-field amplitude – either of these choices can be more adequate. However, one must not mix these gauges. The charm of the electric-field gauge is that one can employ analogies to static electric fields, e.g., tunneling of electrons. In the relativistic regime (see Sect. 4.5), the radiation gauge (Lorentz gauge) should be used. A detailed discussion on these gauge issues can be found in Ref. [89].

(i) Radiation gauge

The first possibility to introduce the laser electric field $\boldsymbol{E}$ into the Schrödinger equation (4.12) is to set $V(x) = -e\phi(x) = 0$. In the same spirit as the dipole approximation in Sect. 3.2, we assume that the relevant lengthscales are much shorter than the wavelength of light, in which case we can approximate the laser electric field as constant in space – but oscillating in time. Suppressing the spatial dependence of $\boldsymbol{E}$ and $\boldsymbol{A}$, we get

$$\boldsymbol{E} = -\frac{\partial \boldsymbol{A}}{\partial t} . \tag{4.15}$$

For an electric field linearly polarized along x and with constant intensity according to $E(t) = \tilde{E}_0 \cos(\omega_0 t + \phi)$, we obtain

$$A_x(t) = -\frac{1}{\omega_0} \tilde{E}_0 \sin(\omega_0 t + \phi) , \tag{4.16}$$

with the CEO phase ϕ. This leads to the time-dependent Schrödinger equation

$$i\hbar \frac{\partial}{\partial t} \psi(x, t) = \frac{1}{2m_e} \left(-i\hbar \frac{\partial}{\partial x} - \frac{e\tilde{E}_0}{\omega_0} \sin(\omega_0 t + \phi) \right)^2 \psi(x, t) . \tag{4.17}$$

Following our above semiclassical discusssion, we make the ansatz

$$\psi(x, t) = e^{ik_x x} \sum_{N=-\infty}^{+\infty} a_N \, e^{-i(\omega_N t + N\phi)} , \tag{4.18}$$

with the frequencies ω_N given by

$$\hbar \omega_N = \frac{\hbar^2 k_x^2}{2m_e} + \langle E_{\mathrm{kin}} \rangle + N \hbar \omega_0 , \tag{4.19}$$

i.e., we have a comb of equidistant frequencies, upshifted by the ponderomotive energy. The term $\hbar^2 k_x^2/(2m_e)$ may be viewed as the initial kinetic energy of the electron. The ansatz (4.18) together with (4.19) can be verified by inserting it into the time-dependent Schrödinger equation (4.17). This, furthermore, delivers the amplitudes a_N. Let us, however, go through the details of the mathematics exclusively for the special case $k_x = 0$ (for $k_x \neq 0$ see Problem 4.2). This leads us to

$$e^{ik_x x} \sum_{N=-\infty}^{+\infty} \hbar \omega_N \, a_N \, e^{-i(\omega_N t + N\phi)} = \frac{1}{2m_e} \frac{e^2 \tilde{E}_0^2}{\omega_0^2} \sin^2(\omega_0 t + \phi) \, \psi(x,t) \quad (4.20)$$

$$= \langle E_{\text{kin}} \rangle \, (1 - \cos(2\omega_0 t + 2\phi)) \, \psi(x,t) \,.$$

Inserting $\psi(x,t)$ according to (4.18) on the RHS, shifting the cosine into the sum, rearranging the sums and comparing the coefficients of the sums gives

$$N \, \hbar \omega_0 \, a_N = -\frac{\langle E_{\text{kin}} \rangle}{2} \, (a_{N+2} + a_{N-2}) \,. \quad (4.21)$$

This relation connects the amplitudes a_N with even integer N to the other even integers, as well as the odd integers to the other odd integers. It does not couple even and odd integers. At this point one needs some inspiration. We guess that

$$a_N = J_{-\frac{N}{2}} \left(\frac{\langle E_{\text{kin}} \rangle}{2 \, \hbar \omega_0} \right) \quad \text{for even integer } N, \quad a_N = 0 \text{ else} \quad (4.22)$$

holds, where J_N is the N-th–order Bessel function of the first kind. Inserting this guess into (4.22), defining $X = \langle E_{\text{kin}} \rangle / (2 \, \hbar \omega_0)$ and replacing $-N/2 = M$, we obtain

$$\frac{2M}{X} J_M(X) = J_{M+1}(X) + J_{M-1}(X) \,, \quad (4.23)$$

a mathematical identity [80] that holds for Bessel functions of the first kind for arbitrary M (integer or half-integer). The amplitude a_N of a given order N depends solely on the ratio of the ponderomotive energy and the carrier photon energy – as already anticipated from our above semiclassical discussion. Without a laser field, i.e., for $\langle E_{\text{kin}} \rangle = 0$, we must recover the usual electron plane waves, i.e., $a_0 = 1$ and all other amplitudes are zero. Indeed, we have $a_0 = J_0(0) = 1$. For nonzero integer indices of the Bessel function, equivalent to even integer order N, we indeed get $J_{-N/2}(0) = 0$. For positive odd integer values of N, equivalent to negative half-integer values of the Bessel function index, on the other hand, $J_{-N/2}(0)$ diverges. Thus, all amplitudes a_N with odd integer value of N must be zero – a consequence of inversion symmetry for $k_x = 0$. At this point, we have verified all aspects of (4.22).

For the more general case $k_x \neq 0$, the amplitudes of the sidebands are given by

$$a_N = e^{+i\frac{\pi}{2} N} \sum_{M=-\infty}^{+\infty} J_M \left(\frac{\langle E_{\text{kin}} \rangle}{2 \, \hbar \omega_0} \right) J_{N-2M} \left(-\frac{k_x \tilde{E}_0 e}{m_e \omega_0^2} \right) \,. \quad (4.24)$$

The argument of the second Bessel function can alternatively be expressed as

$$-\frac{k_x \tilde{E}_0 e}{m_e \omega_0^2} = -\text{sgn}(k_x) \, 2\sqrt{2} \, \frac{\sqrt{\dfrac{\hbar^2 k_x^2}{2m_e} \langle E_{\text{kin}} \rangle}}{\hbar \omega_0} \,, \quad (4.25)$$

i.e., as the geometrical average of the initial kinetic energy and ponderomotive energy divided by the carrier photon energy. For finite values of this ratio, i.e., for finite k_x, Volkov sidebands at even *and* odd orders N occur – in contrast to the case of $k_x = 0$. Note that the sign of a_N for $k_x > 0$ and $k_x < 0$, respectively, is different: The initial electron momentum k_x breaks the inversion symmetry of the free electron. For $k_x = 0$, (4.24) simplifies to (4.22). In order to actually verify (4.24), it is much more convenient to express the laser electric field as a sine rather than as a cosine (this book), because no complex numbers occur in the equation for the coefficients [90]. At the end, one can shift the time axis, which leads to the phase factor $\exp(+i\frac{\pi}{2} N)$ in (4.24). Also, see Problem 4.2.

The states according to (4.18) together with (4.19) and (4.22) or (4.24) are called *Volkov states* [91, 92]. They have a sharp wave number k_x and are completely delocalized in real space: The corresponding charge density $\rho(x, t) \propto -e|\psi(x, t)|^2$ is constant versus x for all times t. The Volkov states are stationary states. They are often refered to as *dressed electron states* because the influence of the light field modifies the "naked" electron dispersion. (4.18) describes exactly those wave packets that we have already discussed in the above semiclassical part. The properties of the Volkov states are visualized in Fig. 4.2.

In more than one dimension and for arbitrary linear polarization of the laser electric field, the term $k_x \tilde{E}_0$ in the argument of the second Bessel function in (4.24) has to be replaced by $\boldsymbol{k} \cdot \tilde{\boldsymbol{E}}_0$ and the wave function $\psi(x, t) \to \psi(x, y, z, t)$ has to be multiplied by the phase factor $e^{i(k_y y + k_z z)}$.

One should note that the Volkov states of vacuum electrons do not introduce *any* nonlinear optics at all. The Volkov states are, however, an important starting point for discussing extreme nonlinear optics of intraband effects in semiconductors or extreme nonlinear optics of atoms. There, a transition into a N-photon sideband of the electron can be viewed as a N-photon absorption process. From the Volkov wave function (4.18) we immediately see that the *corresponding phase is N times the CEO phase ϕ*. Interference of, e.g., one- and two-photon absorption will thus again introduce a dependence on the CEO phase ϕ of the exciting laser pulses.

In the perturbative regime, i.e., for $\frac{\langle E_{\text{kin}} \rangle}{\hbar \omega_0} \ll 1$, we can take advantage of the expansion $J_M(X) \approx (\frac{1}{2} X)^M / \Gamma(M + 1)$ valid for $X \ll 1$ and for positive M. $\Gamma(M + 1)$ is the gamma function with $\Gamma(M + 1) = M!$ for positive integer M. Together with $J_{-M}(X) = (-1)^M J_M(X)$ for integer $M = -N/2$ with (4.22) for $k_x = 0$, we obtain

$$|a_N|^2 \approx \left(\frac{1}{(|N|/2)!}\right)^2 \left(\frac{\langle E_{\text{kin}} \rangle}{4 \hbar \omega_0}\right)^{|N|} \propto I^{|N|}. \tag{4.26}$$

This result could have been anticipated intuitively: The probability to have N photons forming a sideband is proportional to the probability to find one photon to the power of N. This translates into a corresponding scaling with laser intensity I of the N-photon absorption process. Note that $|a_N|^2$ decreases dramatically with increasing $|N|$. For example, for $\langle E_{\text{kin}} \rangle / (\hbar \omega_0) = 0.1 \ll 1$ we get $|a_0|^2 = 1$, $|a_{\pm 2}|^2 = 6.2 \times 10^{-4}$, $|a_{\pm 4}|^2 = 9.8 \times 10^{-8}$, etc.

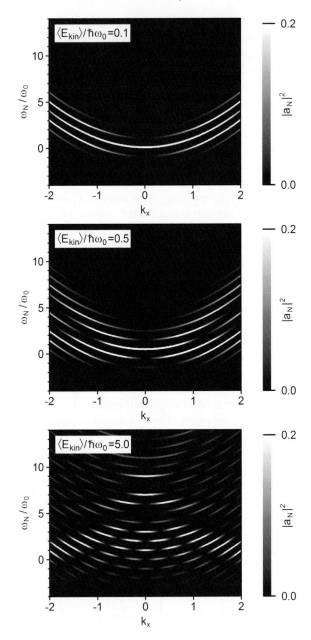

Fig. 4.2. Dispersion relation ω_N versus wave number k_x of Volkov states for a ponderomotive energy of $\frac{\langle E_{\mathrm{kin}}\rangle}{\hbar\omega_0} = 0.1$, $\frac{\langle E_{\mathrm{kin}}\rangle}{\hbar\omega_0} = 0.5$ and $\frac{\langle E_{\mathrm{kin}}\rangle}{\hbar\omega_0} = 5.0$ (as indicated) according to (4.18) and (4.24). The grayscale corresponds to the square modulus of the amplitude, i.e., to $|a_N|^2$, on a linear scale. A normalized wavevector with $k_x^2 = 4$ corresponds to an actual electron kinetic energy of $\frac{\hbar^2 k_x^2}{2m_e} = 4\,\hbar\omega_0$ [268].

Note that the ratio $\frac{\langle E_{\text{kin}} \rangle}{\hbar \omega_0}$ diverges for $\omega_0 \to 0$, i.e., the Volkov states are not useful in electrostatics.

▶ **Example 4.2.** Suppose we are interested in the case that the N-photon sideband of the Volkov states acquires a maximum amplitude $|a_N|$ – obviously a nonperturbative situation. Let us consider $k_x = 0$ and $\hbar \omega_0 = 1.5\,eV$. What is the laser intensity I or ponderomotive energy $\langle E_{\text{kin}} \rangle$ required for this? For even $N \gg 1$, the Bessel function $|J_{-\frac{N}{2}}(X)|$ in (4.22) has its first maximum (the largest maximum) roughly at around $X \approx N/2$. This leads us to $\langle E_{\text{kin}} \rangle \approx N \hbar \omega_0$. For a vacuum electron this translates into a laser intensity of (see (2.16) and (4.6))

$$I = N \times 3 \times 10^{13}\,\text{W/cm}^2 \, ; \quad N \gg 1 \, .$$

This result will become relevant in Sect. 5.3, where we discuss multiphoton ionization of atoms.   ◀

(ii) Electric-field gauge

Choosing $A_x = 0$ as the second option, the laser electric field corresponds to a potential energy $V(x, t) = +x\, e\, E(t)$ of the electron. Thus we have the partial differential equation

$$i\hbar \frac{\partial}{\partial t} \psi(x, t) = -\frac{\hbar^2}{2m_e} \frac{\partial^2}{\partial x^2} \psi(x, t) + x\, e\, \tilde{E}_0 \cos(\omega_0 t + \phi)\, \psi(x, t) \, . \qquad (4.27)$$

The RHS of this special form of the time-dependent Schrödinger equation (4.27) is a quadratic form with time-dependent coefficients. Such a form can generally be solved analytically by means of path integrals [93] for the initial condition $\psi(x, t = 0) = \delta(x)$ (the Green's function or the "propagator" of the problem). This solution is, however, lengthy and not helpful in the context of this book. It is actually simpler to directly numerically solve the Schrödinger equation (4.27). Note that for such a form of the Hamiltonian, the Ehrenfest theorem tells us that *the expectation values $\langle x \rangle(t)$ and $\langle v \rangle(t)$ are strictly identical to the classical observables $x(t)$ and $v(t)$.* An example of additional information provided by quantum mechanics is how the initially localized wave packet disperses and broadens in time.

This is shown in Fig. 4.3 for $\text{Re}(\psi(x, t))$. We chose to depict the real part of the electron wave function here, because it is the counterpart of the electric field of the light, whereas the electron probability density $|\psi(x, t)|^2$ would be the analogue of the light intensity. The simulated region in Fig. 4.3 is significantly larger than the depicted one in order to avoid artificial reflections from the boundaries. The simulation starts at $t = 0$ with a real Gaussian electron wave packet centered at $x = 0$. Its initial velocity is zero. The wave packet is accelerated by the positive laser electric field towards negative x and its center of mass returns to $x = 0$ after one optical cycle at time $t = 2\pi/\omega_0 = 2.8\,\text{fs}$. At this point (not shown), the wave packet has broadened enormously due to group-velocity dispersion. This effect is well known from elementary quantum

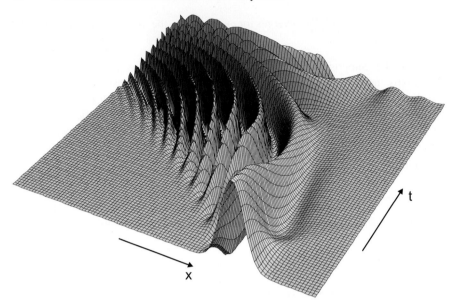

Fig. 4.3. Real part of the electron wave function, $\mathrm{Re}(\psi(x, t))$, obtained from numerical solution of (4.27). Time t runs from 0 to 1.4 fs, the x-coordinate from -3 nm to $+1$ nm. The initial condition at $t = 0$ is $\psi(x, 0) = \exp(-(x/\sigma_0)^2)$, i.e., a resting Gaussian electron wave packet with $\langle x \rangle(0) = 0$ and $\langle v \rangle(0) = 0$. $\sigma_0 = 0.2$ nm corresponds to about four times the hydrogen Bohr radius. The other parameters are: $\phi = 0$, $\hbar\omega_0 = 1.5\,eV$ equivalent to $2\pi/\omega_0 = 2.8$ fs period of light, $m_e = m_0$, and $\tilde{E}_0 = 3 \times 10^{10}$ V/m. This electric field corresponds to a laser intensity of $I = 1.1 \times 10^{14}$ W/cm^2 and a ratio of ponderomotive energy to carrier photon energy of $\frac{\langle E_{kin} \rangle}{\hbar\omega_0} = 5$. Compare with Fig. 4.2.

mechanics: The narrower the initial wave packet in real space, the broader is its momentum distribution according to the uncertainty relation and the more it disperses with time. Even after 1.4 fs (the end of the depicted timescale) corresponding to just half a cycle of light, this broadening is quite prominent. Furthermore, note that the wave packet is strongly "chirped", i.e., its wavelength depends on x and that the phase between a hypothetical envelope of the wave packet and its "carrier" oscillation changes with time t. This is expected from our above semiclassical reasoning. Indeed, for the parameters of Fig. 4.3, the phase changes by $5 \times 2\pi$ for one cycle of light.

Notably, Fig. 4.3 and Fig. 4.2 for $\frac{\langle E_{kin} \rangle}{\hbar\omega_0} = 5$ strictly describe the identical physical situation – in two different gauges as well as in different spaces: One in the Fourier domain, the other one in real space and time.

The static-field limit, i.e., $\omega_0 = 0 = \phi$, is discussed separately in Sect. 7.3. The corresponding electron wave functions are given by Airy functions.

Problem 4.2. Verify (4.24) for the amplitudes a_N of the Volkov states according to (4.18) for the general case of arbitrary initial wave number $k_x \neq 0$.

4.3 Crystal Electrons

For a given intensity of light I, the ponderomotive energy of electrons in semiconductors according to (4.6) is usually much larger than for vacuum electrons, because typical effective electron masses are about an order of magnitude smaller than the free electron mass m_0 (see Table 4.1). For crystal electrons, however, the concept of the ponderomotive energy is only meaningful within the range of validity of the effective-mass approximation, which fails for large values of $\langle E_{kin} \rangle$, often already above several $0.1\,eV$ (see Fig. 7.1). This limits the importance of the ponderomotive energy for optical (i.e., $\hbar\omega_0 = 1.5$ to $3.0\,eV$) excitation of semiconductors under extreme conditions. For infrared excitation with small photon energies $\hbar\omega_0$, on the other hand, one can fulfill the condition $\langle E_{kin} \rangle \approx \hbar\omega_0 < 0.1$ to $0.2\,eV$, in which case the regime of extreme nonlinear "optics" can be reached within the range of validity of the effective-mass approximation. We will come back to corresponding experiments and their description in Sect. 7.4.

Table 4.1. Effective electron mass m_e (in units of the free electron mass $m_0 = 9.1091 \times 10^{-31}$ kg) for a few selected semiconductors. Values taken from Ref. [94].

	GaAs	AlAs	ZnSe	ZnO	ZnTe	CdS	Ge
m_e	0.0665	0.124	0.13	0.24	0.2	0.2	0.0815 ($\perp$), 1.588 ($\parallel$)

The failure of the concept of the ponderomotive energy in solids for large laser intensities calls for a more general quantity that reflects the kinetic energy of the electrons within the bands without employing the effective-mass approximation. Within the acceleration theorem, the crystal-electron momentum, $\hbar k_x$, obeys Newton's second law according to $\hbar \dot{k}_x = F$. Inserting the laser electric field $E(t)$ into the force $F = -eE(t)$, we can easily rewrite the acceleration theorem according to

$$a \frac{\partial}{\partial t} k_x(t) = -\Omega_B(t), \qquad (4.28)$$

where we have introduced the instantaneous *Bloch frequency* $\Omega_B(t)$ with

$$\hbar\Omega_B(t) = a\,e\,E(t). \qquad (4.29)$$

At any given time t, the Bloch energy $\hbar\Omega_B(t)$ is obviously just the potential drop over one unit cell of the crystal lattice with lattice constant a. Note that the Bloch frequency $\Omega_B(t)$ oscillates in time and periodically changes sign in the same way as the Rabi frequency $\Omega_R(t)$ according to (3.18).

4.3.1 Static-Field Case

In order to get an intuitive understanding of the meaning of the Bloch frequency, we consider a static electric field $E(t) = E_0$ [95, 96]. In this case, we can easily solve

(4.28) together with the initial condition $k_x(0) = 0$ and obtain

$$k_x(t) = -t \, \Omega_B/a \,. \tag{4.30}$$

At time $t = \pi/\Omega_B$, the electron hits the end of the first Brillouin zone, i.e., we have $k_x = -\pi/a$. This leads to Bragg reflection of the crystal electron to the other end of the first Brillouin zone with $k_x = +\pi/a$. After another time span of π/Ω_B, the electron is back to its initial state, $k_x = 0$, and one oscillation period $2\pi/\Omega_B$ is completed. This oscillation in wave-number space leads to an oscillation of the electron along the x-direction in real space, which is known as a *Bloch oscillation*. Note that its frequency, the Bloch frequency, does not depend on the particular dispersion relation (the band structure) of the crystal electron.

What is the appropriate quantum-mechanical picture? Without an electric field, the electron wave functions of the atoms forming the solid overlap, which lifts their degeneracy, leading to *delocalized* electron wave functions and bands in the first place. In the presence of a strong electric field, i.e., for $|-a \, e \, E|$ large compared with the width of the band (typically a few electron Volts), the potential drop over one lattice constant, $-a \, e \, E$, lifts the degeneracy and the wave functions become *localized* again. The corresponding eigenenergies, E_M, are evenly separated in energy according to the *Wannier–Stark ladder* [97–99]

$$E_M = M \, a \, e \, E \,, \tag{4.31}$$

with integer $M = -\infty, \, ..., \, -1, \, 0, \, 1, \, +\infty$. An electronic wave packet is a super-position of these Wannier–Stark states and leads to a quantum beating between these states in time. This quantum beating is the quantum-mechanical analogue of the Bloch oscillations. Thus, the *Bloch frequency* Ω_B is given by

$$\hbar\Omega_B = E_{M+1} - E_M = a \, e \, E \,. \tag{4.32}$$

This quantum-mechanical result is identical to that of the semiclassical reasoning, (4.29). Note that (4.31) is analogous to the frequency comb of mode-locked laser oscillators (see Sect. 2.3), where $\Omega_B/2\pi$ plays the role of the repetition frequency f_r (see (2.31)).

4.3.2 High Harmonics from Carrier-Wave Bloch Oscillations

Let us discuss the spectrum of light radiated from a crystal-electron wave packet within a one-dimensional tight-binding band with energy dispersion

$$E_e(k_x) = \hbar\omega_e(k_x) = -\Delta \cos(k_x a) \tag{4.33}$$

(see Fig. 7.1) semiclassically. 2Δ is the width of the band. We consider continuous-wave excitation, i.e., $E(t) = \tilde{E}_0 \cos(\omega_0 t + \phi)$, which is equivalent to

$$\Omega_B(t) = \Omega_B \cos(\omega_0 t + \phi) \tag{4.34}$$

with the *peak Bloch frequency* $\Omega_B = a\,e\tilde{E}_0/\hbar$. The ratio Ω_B/ω_0 is sometimes called the *dynamical localization parameter*. Furthermore, we neglect any type of damping and use the single-particle approximation. The intensity spectrum $I_{rad}(\omega)$ radiated by this wave packet via its intraband motion is proportional to the square modulus of the Fourier transform of the group acceleration, equivalent to $I_{rad}(\omega) \propto |\omega\,v_{group}(\omega)|^2$. The electron group velocity v_{group} at wave number k_x results from

$$v_{group} = \frac{d\omega_e}{dk_x}, \tag{4.35}$$

hence

$$v_{group}(t) = \frac{a\,\Delta}{\hbar}\,\sin(k_x(t)\,a). \tag{4.36}$$

With the initial condition $k_x(0) = 0$ and for CEO phase $\phi = 0$, we obtain the electron wave number $k_x(t)$ from (4.28)

$$k_x(t) = -\frac{\Omega_B}{a\,\omega_0}\,\sin(\omega_0 t). \tag{4.37}$$

Inserting (4.37) into (4.36) leads to

$$v_{group}(t) = -\frac{a\,\Delta}{\hbar}\,\sin\left(\frac{\Omega_B}{\omega_0}\,\sin(\omega_0 t)\right) \tag{4.38}$$

$$= -\frac{a\,\Delta}{\hbar}\,2\sum_{M=0}^{\infty} J_{2M+1}\left(\frac{\Omega_B}{\omega_0}\right)\sin\big((2M+1)\,\omega_0 t\big).$$

In the last step we have employed one of the useful mathematical identities for N-th–order Bessel functions J_N of the first kind (see formula 9.1.43 in Ref. [80]). Finally, from (4.38) the peak heights of the odd harmonics in the radiated intensity spectrum immediately result as

$$I_{rad}(N\omega_0) \propto (N\omega_0)^2\,J_N^2\left(\frac{\Omega_B}{\omega_0}\right). \tag{4.39}$$

The radiated intensity spectrum is illustrated in Fig. 4.4. The zeros of the Bessel functions in (4.39) give rise to the nodes. For $\Omega_B/\omega_0 \ll 1$, the first sine in (4.38) can be approximated by its argument and the group velocity becomes a harmonic oscillation at the carrier frequency ω_0, i.e., $v_{group} \propto -\sin(\omega_0 t)$. This limit recovers our previous result obtained within the effective-mass approximation in Sect. 4.2. For $\Omega_B/\omega_0 \approx 1$, the behavior deviates from this and odd harmonics of ω_0 appear in the Fourier domain via the $\sin(\ldots\sin(\ldots t))$ behavior that stem from the nonparabolicity of the band. For $\Omega_B/\omega_0 \gg 1$ and at around times t with $\omega_0 t = 0, \pi, \ldots$, the second sine can be approximated by its argument, hence the crystal electron harmonically oscillates with the peak Bloch frequency, i.e., $v_{group} \propto \mp\sin(\Omega_B t)$. This represents the static-field limit, which we have already discussed above. As the peak Bloch

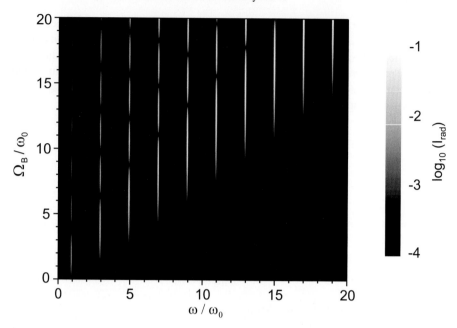

Fig. 4.4. A crystal-electron wave packet in a one-dimensional tight-binding band (see Fig. 7.1) is driven by an intense continuous-wave laser. The radiated intensity spectrum, I_{rad} according to (4.39), is plotted on a logarithmic grayscale versus spectrometer frequency ω and peak Bloch frequency Ω_B in units of the laser carrier frequency ω_0. It is interesting to compare the harmonic spectra for the carrier-wave Bloch oscillations shown here with those of carrier-wave Rabi flopping shown in Fig. 3.11 [269].

frequency is the largest frequency in the system within this limit, it determines the cutoff harmonic order, which is approximately given by

$$N_{cutoff} = \frac{\Omega_B}{\omega_0} .$$

This cutoff (visible as the black "triangle" on the lower RHS in Fig. 4.4) is closely similar to that of the two-level system in the high-field limit (see (3.33) in Sect. 3.6). In contrast to the two-level system, however, only peaks right at spectrometer frequencies $\omega = N \omega_0$ with odd integer N occur (no sidebands, compare Fig. 4.4 with Fig. 3.11).

Apart from the prefactor, the shape of the intensity spectrum $I_{rad}(\omega)$ depends neither on the width of the tight-binding band, 2Δ, nor on the lattice constant a. It is rather solely determined by the dynamical localization parameter Ω_B/ω_0 and is universal in that sense. Note, however, that we have neglected any type of damping (scattering). This approximation becomes questionable when the electron is high in the band, where it has many channels to relax into if Δ is on the order of electron Volts. More advanced theoretical approaches for harmonic generation from band electrons are based on the Boltzmann equation [100–103]. Additionally accounting for scattering within a relaxation time approximation [103] leads to similar results (compare Fig. 3 in Ref. [103] with Fig. 4.4).

Third-harmonic generation from artificial semiconductor superlattices has been observed experimentally [104]. In these experiments, a n-doped GaAs/AlAs super-lattice with $a = 4.85$ nm is excited by the radiation from a free-electron laser at 0.7 THz frequency ($\hbar\omega_0 = 2.9$ meV). Under these conditions, $\Omega_B/\omega_0 = 1 \Leftrightarrow \tilde{E}_0 = 6 \times 10^5$ V/m or $I = 2 \times 10^5$ W/cm^2, both quoted *within* the GaAs using $\epsilon = 10.9$ (see Problem 2.1).

In the optical regime and for real crystals, much larger intensities are required to meet the condition $\Omega_B/\omega_0 = 1$. For example, for $\hbar\omega_0 = 1.5$ eV and the GaAs lattice constant of $a = 0.5$ nm, this condition is equivalent to $\tilde{E}_0 = 3 \times 10^9$ V/m or $I = 4 \times 10^{12}$ W/cm^2, both quoted again *within* the GaAs using $\epsilon = 10.9$. For GaAs, we accidentally have $\Omega_B = \Omega_R$ (compare Example 3.1). We will see in Sect. 7.1 that the condition $\Omega_B/\omega_0 = \Omega_R/\omega_0 \approx 1$ *within* GaAs is just barely compatible with typical GaAs damage thresholds. At this point, nothing special is expected from the carrier-wave Bloch oscillations shown in Fig. 4.4.

In this section, we have only accounted for the *intraband* contribution of the optical polarization. In addition, the intraband driving can also modify the *interband* optical polarization in which case one generally expects a complicated mixture of carrier-wave Rabi oscillations (see Sects. 3.3 and 7.1.4) and carrier-wave Bloch oscillations.

Problem 4.3. Extend our semiclassical discussion of high-harmonic generation to the case of an additional dc-field, such that $\Omega_B(t) = \Omega_B \cos(\omega_0 t) + \Omega_B^{dc}$.

4.4 Extreme Nonlinear Optics of Relativistic Electrons

Let us come back to classical real electrons – but consider the relativistic regime. Our treatment in Sect. 4.1 has accounted for the electric-field component of the light field. But what about the magnetic component? What about the fact that the light field is not constant in space but is rather a wave oscillating in space and time?

▶ **Example 4.3.** For $\tilde{E}_0 = 8 \times 10^{12}$ V/m, $\hbar\omega_0 = 1.5$ eV and free electrons ($m_e = m_0 = 9.1091 \times 10^{-31}$ kg) we get a ponderomotive energy of $\langle E_{kin}\rangle = 540$ keV from (4.6), which is comparable to the relativistic rest energy $m_0 c_0^2 = 512$ keV of the electron. Thus, the nonrelativistic expression of the kinetic energy (4.6) no longer applies. The corresponding intensity is 9×10^{18} W/cm^2. Thus, we anticipate an appreciable influence of relativistic effects already at intensities around 10^{18} W/cm^2. ◀

In order to study these effects, we have to consider a more complete version of Newton's second law for an electron driven by the light field

$$\frac{d(m_e v)}{dt} = F(r,t) = -e\left(E(r,t) + v(t) \times B(r,t)\right), \qquad (4.40)$$

with the laser electric field $E = (E_x, 0, 0)^T$, the laser magnetic field $B = (0, B_y, 0)^T$, wavevector of light $K = (0, 0, K_z)^T$ and

$$E_x(z, t) = \tilde{E}_0 \cos(K_z z - \omega_0 t - \phi) \tag{4.41}$$

$$B_y(z, t) = \tilde{B}_0 \cos(K_z z - \omega_0 t - \phi). \tag{4.42}$$

Here we have assumed a constant field envelope for simplicity and have neglected radiation damping. If the electron velocity v should become relativistic, we furthermore have to account for the relativistic mass

$$m_e = m_e(t) = \frac{m_0}{\sqrt{1 - \dfrac{v^2(t)}{c_0^2}}}. \tag{4.43}$$

$m_0 = 9.1091 \times 10^{-31}$ kg is the electron rest mass.

We have seen in Sect. 4.1 that the electron velocity is proportional to the electric field. Thus, the $v \times B$ term in (4.40) is quadratic in the laser-field amplitude and can be neglected in linear optics. Moreover, the electron does not move along the propagation direction of light in this limit. Hence, its z-coordinate is fixed and we can suppress the z-dependence of the field. Furthermore, for velocities small compared to the speed of light, we can set $m_e = m_0$. With these steps altogether we recover the simple form of Newton's law (4.1). Hence, all results based on this are correct within these limits.

Significant deviations from this behavior are expected if the magnetic component of the Lorentz force becomes comparable to the electric component. Using $\tilde{E}_0/\tilde{B}_0 = c_0$ (see Sect. 2.2), this point is equivalent to the condition $|v_0|/c_0 \approx 1$. With the peak velocity $v_0 = -e\tilde{E}_0/(m_0\omega_0)$ from Newton's second law (with $m_e \to m_0$ and $\tilde{B}_0 \approx 0$), this is, furthermore, equivalent to stating that the *dimensionless parameter* $|\mathcal{E}|$, which is given by

$$\mathcal{E} = \frac{-e\tilde{E}_0}{m_0 \omega_0 c_0}, \tag{4.44}$$

becomes comparable to unity.

If we had just the magnetic-field component of the laser field and if it were a static field, a nonrelativistic electron would simply orbit in circles around the magnetic-field axis with the *cyclotron frequency* ω_c given by

$$\omega_c = \frac{e\tilde{B}_0}{m_0}. \tag{4.45}$$

Introducing (4.45) into (4.44) we get

$$|\mathcal{E}| = \frac{\hbar\omega_c}{\hbar\omega_0}. \tag{4.46}$$

Thus, we can equivalently say that *something special is expected to happen if the cyclotron energy $\hbar\omega_c$ becomes comparable to the carrier photon energy $\hbar\omega_0$.* Yet another equivalent form for $\mathcal{E}$ based on the ponderomotive energy $\langle E_{kin} \rangle$ and the electron rest energy $m_0\,c_0^2$ is given in (4.76).

▶ **Example 4.4.** In vacuum, for electrons with rest mass $m_0 = 9.1091 \times 10^{-31}$ kg, carrier photon energy $\hbar\omega_0 = 1.5\,eV$ and with the fundamental constants e and c_0, we have

$$|\mathcal{E}| = 1 \tag{4.47}$$

$$\Leftrightarrow$$

$$\tilde{E}_0 = 3.9 \times 10^{12}\,\text{V/m}$$

$$\Leftrightarrow$$

$$\tilde{B}_0 = 1.3 \times 10^4\,\text{T}$$

$$\Leftrightarrow$$

$$I = 1.9 \times 10^{18}\,\text{W/cm}^2\,.$$

◀

To get rid of the various constants and to make the mathematics more transparent, we employ the normalized electric-field strength $\mathcal{E}$ and, furthermore, introduce the normalized coordinates $\tilde{x} = x\,\omega_0/c_0$ and $\tilde{z} = z\,\omega_0/c_0$, normalized dimensionless time $\tilde{t} = t\,\omega_0$, as well as the usual relativistic parameters $\boldsymbol{\beta} = \boldsymbol{v}/c_0$ and

$$\gamma = \frac{m_e}{m_0} = \frac{1}{\sqrt{1 - \left(\dfrac{v}{c_0}\right)^2}} = \frac{1}{\sqrt{1 - \beta_x^2 - \beta_y^2 - \beta_z^2}}\,. \tag{4.48}$$

It is then straightforward to rewrite Newton's law (4.40) for CEO phase $\phi = 0$ in components as

$$\frac{\mathrm{d}(\gamma\beta_x)}{\mathrm{d}\tilde{t}} = \mathcal{E}\,(1 - \beta_z)\,\cos(\tilde{z} - \tilde{t}) \tag{4.49}$$

and

$$\frac{\mathrm{d}(\gamma\beta_z)}{\mathrm{d}\tilde{t}} = \mathcal{E}\,\beta_x\,\cos(\tilde{z} - \tilde{t})\,. \tag{4.50}$$

The force along y is zero.

4.4.1 Second-Harmonic Generation and Photon Drag

We will discuss the complete problem intuitively and mathematically in a moment. It is instructive, however, to first consider a perturbative approach, which leads us to two interesting effects already: The photon drag and second-harmonic generation on free electrons. As discussed above, for low intensities, we have $\gamma \approx 1$, $\beta_z \ll 1$ and $\tilde{z} \ll \tilde{t}$, in which case (4.49) reduces to

$$\frac{d\beta_x}{d\tilde{t}} = \mathcal{E} \cos(\tilde{t}). \tag{4.51}$$

With the initial conditions $\beta_x(0) = 0$ and $\tilde{x}(0) = 0$, the solution is

$$\beta_x(\tilde{t}) = \mathcal{E} \sin(\tilde{t}) \quad \text{and} \quad \tilde{x}(\tilde{t}) = \mathcal{E}\left(1 - \cos(\tilde{t})\right). \tag{4.52}$$

Introducing this expression for β_x on the RHS of (4.50) (again with $\gamma \approx 1$ and $\tilde{z} \ll \tilde{t}$) in the spirit of a perturbation expansion leads to

$$\frac{d\beta_z}{d\tilde{t}} = \mathcal{E}^2 \sin(\tilde{t}) \cos(\tilde{t}) = \mathcal{E}^2 \frac{1}{2} \sin(2\tilde{t}). \tag{4.53}$$

With the initial conditions $\beta_z(0) = 0$ and $\tilde{z}(0) = 0$, the solution is

$$\beta_z(\tilde{t}) = \frac{\mathcal{E}^2}{4}\left(1 - \cos(2\tilde{t})\right) \quad \text{and} \quad \tilde{z}(\tilde{t}) = \frac{\mathcal{E}^2}{4}\left(\tilde{t} - \frac{1}{2}\sin(2\tilde{t})\right). \tag{4.54}$$

The plots of (4.52) and (4.54) for $\mathcal{E} = 0.1$ are indistinguishable (within the line thickness) from those of the exact analytical calculation, which are shown on the LHS and RHS of Fig. 4.6(a), respectively. It can be seen that the motion along the z-direction contains two contributions: (i) A constant drift and (ii) an oscillation with twice the laser carrier frequency. Let us have a closer look at both of them.

(i) The drift motion has a velocity $\langle \beta_z \rangle = \mathcal{E}^2/4$ (0.25% of the speed of light in Fig. 4.6(a)) and is directed along the propagation direction of light. It corresponds to a constant electrical current density, the so-called *photon-drag current* j_{pd}, which is proportional to the light intensity ($I \propto \tilde{E}_0^2 \propto \mathcal{E}^2$) – "the photons push the electron". Translating back into physical units, the photon-drag current density under these conditions is given by

$$j_{pd} = \frac{-eN_e}{V} c_0 \langle \beta_z \rangle = -\frac{N_e \, e^3}{4 \, V m_0^2 \, c_0 \, \omega_0^2} \tilde{E}_0^2. \tag{4.55}$$

Note that the sign of this unusual current depends on the sign of the charge of the particle because j_{pd} is proportional to the cube of the charge. For example, in an isotropic semiconductor, the hole photon-drag current would be opposite to the electron contribution to the current. Both electrons and holes are, however, pushed into the direction of the wavevector of light. The current $j_{pd} \propto 1/\omega_0^2$ obviously increases with decreasing photon energy $\hbar\omega_0$. This is the basis for using the photon-drag current in commercially available semiconductor *infrared* photodetectors, see Sect. 7.4.

One might be tempted to argue that a constant light intensity cannot accelerate an electron as the RHS of (4.50) and (4.53) do not contain any dc component under these conditions. This is indeed true. So how does the electron acquire its drift velocity? It is accelerated when the light intensity is ramped up. The exact value of the drift velocity does depend on the way the light intensity is ramped up, i.e., it is not a unique function of the instantaneous light intensity. Our result, (4.54), corresponds to particular initial conditions, namely $r(0) = v(0) = 0$ (also see Problem 4.4). When the light intensity in the pulse decreases in time, the electron slows down again.

(ii) The second component of the motion is an oscillation of the electron displacement $z(t)$, hence also of the optical polarization $P_z(t)$, with twice the laser carrier frequency, i.e., $z(t) \propto P_z(t) \propto \sin(2\omega_0 t)$ when translated back to physical units. This is simply *second-harmonic generation* – despite the fact that an electron in vacuum has inversion symmetry. Note that the electron oscillation is along the z-direction, i.e., it is perpendicular to E as well as perpendicular to B, thus parallel to the wavevector of light K. Lumping all prefactors together in a second-order susceptibility $\chi_L^{(2)}$, we indeed get (2.43). There, we have already discussed the symmetry properties of this contribution with respect to space inversion. The corresponding radiation pattern of the second harmonic from a single electron is illustrated in Fig. 4.5(b). The corresponding trajectory of the electron is a *figure-of-eight* motion in the xz plane (see Fig. 4.5) because it oscillates with frequency ω_0 along x and with frequency $2\omega_0$ along z.

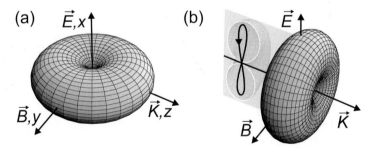

Fig. 4.5. (a) Radiation pattern according to linear Thomson scattering (carrier frequency ω_0) on free electrons. **(b)** Corresponding second-order nonlinear Thomson scattering (not to scale) with carrier frequency $2\omega_0$ in the perturbative regime ($\mathcal{E}^2 \ll 1$). The vectors of the electric field E, the magnetic field B and the wavevector of light K are also depicted. Note that no second harmonic is emitted into the propagation direction of light. The center shows the figure-of-eight trajectory of the electron in the xz-plane in a frame moving with the electron drift velocity along the z-direction (photon drag) [269].

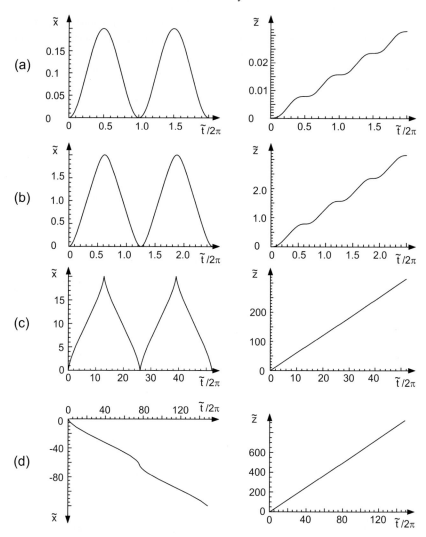

Fig. 4.6. Relativistic motion (laboratory frame) of a free electron in vacuum under the influence of a strong laser field according to (4.56)–(4.58). The light propagates along z with $\mathbf{E}$ and $\mathbf{B}$ being polarized along x and y, respectively. $\mathcal{E}$ is the normalized electric field strength given by $|\mathcal{E}| = \omega_c/\omega_0$. (a) $\mathcal{E} = 0.1$, $\zeta_0 = 0$; (b) $\mathcal{E} = 1$, $\zeta_0 = 0$; (c) $\mathcal{E} = 10$, $\zeta_0 = 0$; and (d) $\mathcal{E} = 10$, $\zeta_0 = \pi/2$. Note the different vertical and horizontal scales. In (a), the oscillation period of $\tilde{z}(\tilde{t})$ is π, equivalent to a motion of $z(t)$ with frequency $2\omega_0$, which leads to second-harmonic generation. In (b) and (c), the period of $\tilde{x}(\tilde{t})$ becomes larger than 2π, equivalent to an oscillation frequency of $x(t)$ smaller than ω_0 due to the relativistic Doppler redshift. Indeed, $\tilde{z}(\tilde{t}) \approx \tilde{t}$ in (c) is equivalent to a drift velocity along z close to the speed of light c_0. The trajectories as well as the periods also depend on the initial condition of the electron. For example, the period of $\tilde{x}(\tilde{t})$ in (c) is $26 \times 2\pi$, while it is $76 \times 2\pi$ in (d) – although the laser intensity is the same [267].

Problem 4.4. In semiconductors or metals, the situation is different from vacuum electrons because crystal electrons are usually subject to scattering, which can be described by adding a Stokes-damping term in Newton's law to a first approximation. In this case, the steady-state photon-drag current does *not* depend on the initial conditions – in contrast to what we have found for vacuum electrons. Show this.

4.4.2 Nonperturbative Regime

We could in principle continue along these lines of perturbation theory (valid for $\mathcal{E}^2 \ll 1$) and insert the expression for β_z into (4.49), which leads to third-harmonic generation with a polarization along x, which then leads to fourth-harmonic generation polarized along z, etc.

Let us rather discuss the exact nonperturbative solution, which allows us to address laser intensities approaching the condition $\mathcal{E}^2 \approx 1$ or even $\mathcal{E}^2 \gg 1$. This leads us to *relativistic nonlinear Thomson scattering* on free electrons (theory: [105–109], experiment: [110]) – sometimes also referred to as *Larmor radiation*. As the light intensity increases, the drift velocity along z also increases, at some point becoming comparable to the vacuum speed of light c_0. Three aspects become important as a result of this: (i) The spatial dependence of the incident fields, i.e., $E_x(z, t) = \tilde{E}_0 \cos(K_z z - \omega_0 t - \phi)$ and $B_y(z, t) = \tilde{B}_0 \cos(K_z z - \omega_0 t - \phi)$, has to be accounted for – "the electron rides on the electromagnetic wave like a surfer". (ii) The laser frequency becomes Doppler redshifted from the perspective of the electron, and (iii) the relativistic mass according to (4.43) changes with time. Let us have a closer look at (i)–(iii). Aspect (iii) is certainly an additional source of optical nonlinearities, hence a source of high harmonics. It is loosely related to the nonparabolicity of the dispersion relation of crystal electrons discussed in Sect. 4.3.2. Aspects (i) and (ii) are related. For relativistic electron velocities, the "source" of the incident electromagnetic wave (the laser) and the "observer" (the electron) move away from each other due to the relativistic drift velocity of the electron (which is parallel to the light wavevector). This leads to a relativistic Doppler redshift. Thus, the electron "feels" a driving frequency that is smaller than ω_0. Note, however, that this frequency cannot simply be calculated from the usual textbook (longitudinal) Doppler effect formulae that would only apply if the electron was a system of inertia – which it is not. In the "surfer picture", the electron rides along with the wave and "moves up and down" slower than a fictitious electron fixed at some position z.

To see the details, we have to solve the relativistic version of Newton's law. At first sight, it seems hopeless to solve (4.40) with the relativistic mass (4.43) under these conditions exactly. Nevertheless, and quite amazingly, for a plane wave with a constant light intensity, an exact analytical implicit solution can be given in terms of a parameter ζ. Assuming a CEO phase $\phi = 0$ and using the normalized coordinates and time introduced above, one gets [108, 111]

$$\tilde{x}(\zeta) = \mathcal{E}\left((\cos\zeta_0 - \cos\zeta) - (\zeta - \zeta_0)\sin\zeta_0\right), \tag{4.56}$$

$$\tilde{z}(\zeta) = \tilde{t} - \zeta, \tag{4.57}$$

$$\tilde{t}(\zeta) = (\zeta - \zeta_0)\left[1 + \frac{\mathcal{E}^2}{2}\left(\frac{1}{2} + \sin^2\zeta_0\right)\right]$$

$$+ \frac{\mathcal{E}^2}{2}\left[-\frac{\sin(2\zeta)}{4} + 2\cos\zeta\,\sin\zeta_0 - \frac{3\sin(2\zeta_0)}{4}\right]. \qquad (4.58)$$

Here, the initial electron velocity v is assumed to be zero at time $t = 0$. The parameter ζ_0 in (4.56) and (4.58) results from the initial position of the electron at $t = 0$ and is given by $\tilde{z}(\tilde{t} = 0) = -\zeta_0$ (see (4.57)). It can be interpreted as the *initial phase of the electron* and is *equal to the CEO phase* ϕ *at this point*, but not identical to it in general: For a cloud of electrons with an extent along the z-direction comparable to or even larger than the wavelength of light $2\pi c_0/\omega_0$, one has a distribution of values for ζ_0. A similar effect can occur in the actual focus of a lens where the phase fronts are not plane everywhere. Selected examples of electron trajectories are given in Fig. 4.6. In each of the plots, ζ runs from 0 to 4π.

In the limit $\mathcal{E}^2 \ll 1$ (see Fig. 4.6(a)), one has $\tilde{t} = \zeta - \zeta_0$, thus $\tilde{z} = -\zeta_0 = \text{const.}$, while the x-coordinate oscillates harmonically with time, i.e., $x(t) \propto \cos(\omega_0 t)$ – as explained above. In Fig. 4.6(b) where $\mathcal{E} = 1$, the excursion along the x-direction is already on the order of $\tilde{x} = 1$, corresponding to an actual value of about $x = 0.1\,\mu\text{m}$ for $\hbar\omega_0 = 1.5\,eV$. In order to make our model of isolated electrons in vacuum realistic, residual scatterers (e.g. ionized atoms) in the vacuum should have a number N_{atom} per volume V of less than $(1/x)^3 = 10^{15}\,\text{cm}^{-3} = 10^{21}\,\text{m}^{-3}$. With the equation of state of ideal gases from thermodynamics, i.e., $PV = N_{\text{atom}}k_B T$ with Boltzmann's constant $k_B = 1.3804 \times 10^{-23}\,\text{J/K}$, this density translates into a maximum residual pressure of $P = 10^{21}\,\text{m}^{-3}k_B T = 4\,\text{Pa}$ at $T = 300\,\text{K}$, which is easy to achieve.

Deep in the relativistic regime, i.e., for $\mathcal{E}^2 = 10^2 \gg 1$ in Fig. 4.6(c), the x-coordinate versus time (LHS of Fig. 4.6(c)) exhibits pronounced sharp maxima. They arise as a result of the fact that – at these points – the oscillatory part of the motion along z is opposite to the drift motion along z, thus β_z is relatively small (it is even zero at these points for $\mathcal{E}^2 \ll 1$, see above perturbative approach). Moreover, the x-component of the velocity, β_x, is strictly zero right at these maxima. Consequently, the relativistic factor $\gamma = \gamma(\beta_x(t), \beta_z(t))$ is closer to 1, the electron becomes "light", hence the electron acceleration (the curvature of the displacement versus time) is large – and a sharp tip in $x(t)$ results. Another striking aspect is that the electron oscillation frequency, ω_0^e, is smaller than ω_0 by a factor of 26 in Fig. 4.6(c) – as anticipated from our above qualitative reasoning based on the Doppler effect. Mathematically, ω_0^e can easily be derived from (4.58). For one oscillation cycle, the parameter ζ increases by 2π. Thus, the normalized time $\tilde{t}$ according to (4.58) changes by $\Delta\tilde{t} = 2\pi(1 + \frac{\mathcal{E}^2}{2}(\frac{1}{2} + \sin^2\zeta_0))$, hence the real time $t = \tilde{t}/\omega_0$ by the period $\Delta t = \Delta\tilde{t}/\omega_0$. The *electron oscillation frequency* $\omega_0^e = 2\pi/\Delta t$ results as

$$\frac{\omega_0^e}{\omega_0} = \frac{1}{1 + \dfrac{\mathcal{E}^2}{2}\left(\dfrac{1}{2} + \sin^2\zeta_0\right)} \leq 1. \qquad (4.59)$$

For example, for $\mathcal{E} = 10$ and $\zeta_0 = 0$, corresponding to Fig. 4.6(c), we obtain a factor of $1/26$ from (4.59). Note that the electron oscillation frequency as well as the shape of the electron trajectory not only depend on the laser intensity via the normalized field strength $|\mathcal{E}|$ but generally also on the initial condition ζ_0, as exemplified in Figs. 4.6(c) and (d). Here $\mathcal{E} = 10$ is fixed and ζ_0 is varied from $\zeta_0 = 0$ in (c) to $\zeta_0 = \pi/2$ in (d). With $\mathcal{E} = 10$ and $\zeta_0 = \pi/2$, (4.59) indeed delivers a factor of $1/76$.

The acceleration of the charged electron is the source of electromagnetic radiation. As the electron motion is periodic (apart from the constant drift velocity) but not harmonic at all, we expect that the *intensity spectrum* $I(\omega)$ – as detected with a spectrometer fixed in the laboratory frame – consists of a series of equidistant peaks, separated by $\tilde{\omega}_0$, i.e.,

$$I(\omega) = \sum_{N=1}^{\infty} I_N\, \delta(\omega - N\,\tilde{\omega}_0),\qquad (4.60)$$

where the order N covers *odd and even harmonics* of the fundamental emission frequency $\tilde{\omega}_0$. Figure 4.7 schematically illustrates the intensity spectrum for some detection direction. *In general, $\tilde{\omega}_0$ is neither identical to the laser carrier frequency ω_0 nor identical to the electron oscillation frequency ω_0^e.* In order to evaluate the fundamental emission frequency $\tilde{\omega}_0$, one has to consider the relativistic Doppler effect one more time: Suppose that we detect light in a backscattering geometry, i.e., in the $-z$ direction. During the electron oscillation, the electron (the "source") moves along z, hence away from the spectrometer (the "observer"). This leads to a Doppler redshift with respect to the redshifted electron oscillation frequency. In contrast to this, for detection in the forward direction, the electron moves towards the "observer", thus we anticipate a blueshift of the electron oscillation frequency – which itself is redshifted with respect to the laser carrier frequency ω_0.

Mathematically, the fundamental emission frequency $\tilde{\omega}_0$ can be related to ω_0 via the following reasoning: The detection direction is along the vector $\tilde{K}_0$, which includes an angle θ with the z-axis (propagation direction of the laser). The modulus

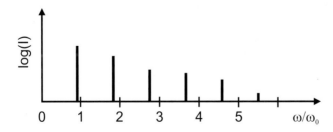

Fig. 4.7. Scheme of the intensity spectrum $I(\omega)$ of light emitted by an electron in vacuum driven by a strong laser field at carrier frequency ω_0 according to (4.60) with $\tilde{\omega}_0/\omega_0 \approx 0.9$ corresponding to $|\mathcal{E}| \approx 0.6$ and $\zeta_0 = 0$. Note that the ratio $\tilde{\omega}_0/\omega_0$ depends on the intensity $\mathcal{E}^2$, on the electron initial phase ζ_0 (see Fig. 4.6) as well as on the detection direction.

of $\tilde{K}_0$ be the fundamental emission wave number of light, following the dispersion relation $\tilde{\omega}_0/|\tilde{K}_0| = c_0$. During one electron oscillation period Δt (see above), the parameter ζ in (4.56) and (4.57) increases by 2π and the electron experiences a displacement Δr. This displacement is given by

$$\Delta r = \begin{pmatrix} \Delta x \\ \Delta y \\ \Delta z \end{pmatrix} = \frac{c_0}{\omega_0} \begin{pmatrix} -2\pi \, \mathcal{E} \sin\zeta_0 \\ 0 \\ 2\pi \, \dfrac{\mathcal{E}^2}{2} \left(\dfrac{1}{2} + \sin^2\zeta_0 \right) \end{pmatrix}. \tag{4.61}$$

For constructive interference of the emission from one cycle and that of the next (and for all subsequent ones), the condition

$$|\tilde{K}_0 \cdot \Delta r - \tilde{\omega}_0 \Delta t| = 2\pi \tag{4.62}$$

has to be fulfilled. $\tilde{\omega}_0$ immediately follows from this condition for any value of ζ_0. For the special case $\zeta_0 = 0$, where $\Delta x = 0$, we obtain the simple formula for the *fundamental emission frequency* $\tilde{\omega}_0$

$$\frac{\tilde{\omega}_0}{\omega_0} = \frac{1}{1 + \dfrac{\mathcal{E}^2}{4}(1 - \cos\theta)} \leq 1. \tag{4.63}$$

For forward scattering ($\theta = 0$), this leads to $\tilde{\omega}_0/\omega_0 = 1$, for backscattering ($\theta = \pi$) to $\tilde{\omega}_0/\omega_0 = 1/(1+\mathcal{E}^2/2)$ – as expected from our above qualitative reasoning. Detection perpendicular to the laser beam ($\theta = \pm\pi/2$) gives $\tilde{\omega}_0/\omega_0 = \omega_0^e/\omega_0 = 1/(1 + \mathcal{E}^2/4)$. Generally, the fundamental emission frequency $\tilde{\omega}_0$ additionally depends on the electron phase ζ_0 (and/or the CEO phase ϕ).

The peak heights I_N also depend on $\mathcal{E}^2$, ζ_0 and on the detection direction. Generally, they can be evaluated numerically based on the usual relativistic formula for the far-field emission of an accelerated charge from electromagnetism [114] via the Lienard–Wichert potentials. This gives rise to complicated radiation patterns, which no longer resemble those shown in Fig. 4.5 for $N = 1$ and $N = 2$. A simple analytical approximate expression for the coefficients I_N has been derived [112, 113] in the high-field limit, i.e., for $|\mathcal{E}| \gg 1$, for $\zeta_0 = 0$ and for detection in the backscattering direction. Here, the intensities I_N for odd N are given by

$$I_N \propto \mathcal{E}^4 \frac{N}{N_{\max}} \exp\left(-\frac{N - N_{\max}}{N_{\max}} \right). \tag{4.64}$$

This dependence starts with a linear increase in the harmonic order N for $N \ll N_{\max}$, followed by a maximum at $N = N_{\max}$ given by

$$N_{\max} \approx 0.32 \times |\mathcal{E}|^3. \tag{4.65}$$

For N well above $N_{\max}$, I_N decays exponentially with N. For example, for $\mathcal{E} = 10$ (see Fig. 4.6(c)), we have $N_{\max} = 320$ and $\tilde{\omega}_0/\omega_0 = 1/51$ from (4.63). This translates

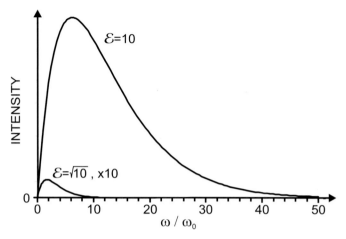

Fig. 4.8. Nonlinear Thomson backscattering intensity (linear scale) versus spectrometer frequency ω in units of the laser carrier frequency ω_0 from (4.60), (4.64) and (4.65). $\zeta_0 = 0$; $\mathcal{E} = \sqrt{10}$ and $\mathcal{E} = 10$ as indicated. For clarity, the curve for $\mathcal{E} = \sqrt{10}$ has been stretched vertically by a factor of ten. $\mathcal{E} = 10$ corresponds to an actual laser intensity of 1.9×10^{20} W/cm^2 for $\hbar\omega_0 = 1.5\,eV$ [269].

into a maximum at spectrometer frequency $\omega = \frac{320}{51}\,\omega_0 \approx 6\,\omega_0$. It is more than likely that this large number of densely spaced harmonics will merge into a continuous spectrum in an actual experiment. This is illustrated in Fig. 4.8. Note that no sharp cutoff in the harmonic order N occurs, whereas a cutoff has been discussed above for high harmonics from two-level systems in Sect. 3.6.1 and will again appear below for high-harmonic generation from atoms in Sect. 5.4.1 (also see Problem 4.3).

Finally, remember that we have discussed the "very simple" case of a light field corresponding to a plane wave with constant intensity in time, interacting with an isolated electron initially at rest. An interesting extension is the case of finite or even relativistic initial electron velocity (see Sect. 8.2). For pulsed excitation, $\mathcal{E}$ and thus also $\tilde{\omega}_0$ are expected to vary in time, the transverse (and longitudinal) beam profile in the focal spot of a lens leads to a dependence on the spatial coordinate. All these aspects modify the nonlinear Thomson scattering spectra.

Problem 4.5. Consider second-harmonic generation from relativistic nonlinear Thomson scattering for continuous-wave excitation with $\mathcal{E}^2 = 1$ and $\zeta_0 = 0$. We have seen that the SHG tends to be redshifted with respect to $2\omega_0$. Hence, we detect the SHG with a filter or a spectrometer in a *narrow* spectral interval centered around, e.g., $0.985 \times 2\omega_0$. What is the shape of the emission pattern that you expect to detect?

4.5 Extreme Nonlinear Optics of Dirac Electrons

We have to distinguish between a classical and a quantum-mechanical treatment of the electron on the one hand, and between nonrelativistic and relativistic behavior on the other hand. Table 4.2 refers to those three cases that we have already discussed in this chapter. The fourth entry is the (special) relativistic quantum-mechanical case, which we want to discuss in what follows.

Table 4.2. A vacuum electron interacts with a light field of constant intensity. Overview about the cases discussed in this chapter. The specifics of crystal electrons are discussed in Sect. 4.3. Effects of general relativity will be touched on in Sect. 4.6

	Classical	Quantum mechanical
Nonrelativistic	Sect. 4.1	Sect. 4.2
Relativistic	Sect. 4.4	this section

It was one of Einstein's discoveries that the energy E of a vacuum electron with momentum p in the relativistic regime is given by the relation

$$E^2 = (m_0 c_0^2)^2 + (p\, c_0)^2 \,. \tag{4.66}$$

It is then usually argued that $E = +\sqrt{\text{RHS}}$, which leads to his famous result

$$E = +\sqrt{(m_0 c_0^2)^2 + (p\, c_0)^2} = +m_e c_0^2 \,. \tag{4.67}$$

m_e is the relativistic electron mass according to (4.43). For small momenta p, Taylor expansion immediately leads us back to the classical kinetic energy

$$E_{\text{kin}} = E - m_0 c_0^2 = \frac{p^2}{2m_0} = \frac{m_0}{2} v^2 \,, \tag{4.68}$$

with the electron rest energy $m_0 c_0^2$. The mathematically possible alternative solution

$$E = -\sqrt{(m_0 c_0^2)^2 + (p\, c_0)^2} = -m_e c_0^2 \tag{4.69}$$

has to be discarded as in classical relativistic physics electrons could "disappear" towards minus infinity in energy, which is considered to be unphysical. Both branches ("+" and "−") are visualized in Fig. 4.9.

Note that our discussion on nonlinear Thomson scattering in Sect. 4.4 referred to the dynamics of an electron in the upper branch subject to the laser field. The lower branch did not occur at all.

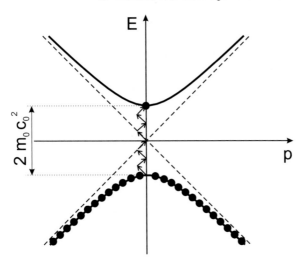

Fig. 4.9. Relativistic quantum-mechanical dispersion relation of electrons in vacuum, i.e., $E = \pm\sqrt{(m_0c_0^2)^2 + (p\,c_0)^2}$, and dispersion relation of photons (dashed straight lines with $E = \pm p\,c_0$). The vacuum corresponds to a fully occupied lower band and an empty upper band as indicated by the dots. An unoccupied state in the lower band is called a positron, an occupied state in the upper band is called an electron. Compare with Fig. 7.1. The arrows indicate the generation of an electron–positron pair by many photons from two counterpropagating light beams (not to scale). Actually, around 10^6 laser photons with $\hbar\omega_0 \approx 1\,eV$ are necessary as $2\,m_0c_0^2 = 1.024\,MeV$ [267].

When P. M. Dirac quantized relativistic mechanics some years later, the exact same problem re-appeared. In quantum mechanics, however, these negative energy states are less problematic and Dirac courageously postulated in 1930 that all of the negative energy states are occupied (this is an infinite number even if the universe is finite in size) – the so-called *Dirac sea* (see dots in Fig. 4.9). The Pauli exclusion principle then guarantees that normal electrons – which are the occupied states on the upper branch – do not disappear. The vacuum is not empty – on the contrary, it is half full. As long as the electrons on the lower branch stay in their branch, they remain almost undetectable to the observer. But Dirac also recognized that one could promote an electron from the lower branch into the upper branch by providing a minimum energy of $2\,m_0c_0^2 = 1.024\,MeV$. In this case, an electron in the upper branch and a missing electron in the lower branch would be generated – an electron–positron pair. This theoretical prediction eventually led to the discovery of the positron, the antiparticle partner of the electron.

Can a single photon corresponding to a plane wave generate an electron–positron pair? In such a process, momentum and energy conservation have to be fulfilled simultaneously. The dispersion relation of the photon $\omega/|\boldsymbol{K}| = c_0$ is equivalent to

$$E = p\,c_0\,, \tag{4.70}$$

which is also depicted in Fig. 4.9 (dashed diagonal lines). It is obvious that one can shift the photon dispersion upwards and/or sideways, but one can never get it to cross with both the lower and the upper electron branch – the photon momentum is too large. This problem can be circumvented by using counterpropagating laser beams with an effective photon momentum equal to zero (or by considering other four-particle processes). Such a "zigzag" multiphoton transition is also depicted in Fig. 4.9.

Mathematically, we have to solve the *Dirac equation* for vacuum electrons. With potential $\phi(\mathbf{r}, t)$ and vector potential $\mathbf{A} = (A_1, A_2, A_3)^T = (A_x, A_y, A_z)^T$ it is given by

$$i\hbar \frac{\partial}{\partial t} \psi(\mathbf{r}, t) = \left(\hat{\beta} m_0 c_0^2 - e\phi(\mathbf{r}, t) + \sum_{n=1}^{3} c_0 \hat{\alpha}_n \left(-i\hbar \frac{\partial}{\partial x_n} + eA_n(\mathbf{r}, t) \right) \right) \psi(\mathbf{r}, t).$$
(4.71)

The (4×4) Dirac matrices $\hat{\alpha}_n$ and $\hat{\beta}$ can be expressed as

$$\hat{\alpha}_1 = \begin{pmatrix} 0 & 0 & 0 & +1 \\ 0 & 0 & +1 & 0 \\ 0 & +1 & 0 & 0 \\ +1 & 0 & 0 & 0 \end{pmatrix}, \quad \hat{\alpha}_2 = \begin{pmatrix} 0 & 0 & 0 & +i \\ 0 & 0 & -i & 0 \\ 0 & +i & 0 & 0 \\ -i & 0 & 0 & 0 \end{pmatrix},$$
(4.72)

$$\hat{\alpha}_3 = \begin{pmatrix} 0 & 0 & +1 & 0 \\ 0 & 0 & 0 & -1 \\ +1 & 0 & 0 & 0 \\ 0 & -1 & 0 & 0 \end{pmatrix} \quad \text{and} \quad \hat{\beta} = \begin{pmatrix} +1 & 0 & 0 & 0 \\ 0 & +1 & 0 & 0 \\ 0 & 0 & -1 & 0 \\ 0 & 0 & 0 & -1 \end{pmatrix}.$$

$\psi = (\psi_1, \psi_2, \psi_3, \psi_4)^T$ is a 4-vector, representing the four possibilities: electron spin up/down and upper/lower band, respectively.

Finding solutions of the Dirac equation under realistic conditions is not a simple task at all. Notably, Volkov recognized in 1935 [91] that the Dirac equation can be solved exactly in the field of an electromagnetic plane wave. (The nonrelativistic limit of this result, the Volkov states, has been discussed in Sect. 4.2.) Here, we only consider the very simple although instructive static limit. We have encountered this "static-field approximation" previously in the context of two-level systems (see Sect. 3.6.1). Suppose we have $\phi(\mathbf{r}, t) = 0$ (radiation gauge, Sect. 4.2) and that the vector potential corresponding to the standing-wave pattern of two counter-propagating (along z) laser beams in an antinode is given by $\mathbf{A} = \tilde{A}_0 (1, 0, 0)^T$ at some instant in time. As the photon drag discussed in Sect. 4.4 is absent here, we can assume that the electron momentum is zero, i.e., the spatial derivatives in (4.71) vanish at this point in time. Under these conditions, ψ_1 couples only to ψ_4 (and ψ_2 only to ψ_3) and the Dirac equation (4.71) simplifies to

$$i\hbar \frac{\partial}{\partial t} \begin{pmatrix} \psi_1 \\ \psi_4 \end{pmatrix} = \begin{pmatrix} +m_0 c_0^2 & c_0 e \tilde{A}_0 \\ c_0 e \tilde{A}_0 & -m_0 c_0^2 \end{pmatrix} \begin{pmatrix} \psi_1 \\ \psi_4 \end{pmatrix}.$$
(4.73)

and a similar form for the 23-subspace.

This immediately reminds one of the two-level system physics discussed in Sect. 3.2. Indeed, in matrix form (see (3.50)) it can be expressed as

$$i\hbar \frac{\partial}{\partial t} \begin{pmatrix} a_2 \\ a_1 \end{pmatrix} = \begin{pmatrix} E_2 & -\hbar\Omega_R \\ -\hbar\Omega_R & E_1 \end{pmatrix} \begin{pmatrix} a_2 \\ a_1 \end{pmatrix}. \tag{4.74}$$

The transition energy is $\hbar\Omega = (E_2 - E_1)$. We have seen in Sect. 3.6.1 that the "static-field approximation" is justified for $\hbar\Omega_R \gg \hbar\omega_0$ and Rabi oscillations could even occur for $\hbar\Omega \gg \hbar\omega_0$ if $\hbar\Omega_R \approx \hbar\Omega$ or $\hbar\Omega_R \gg \hbar\Omega$.

In analogy to the two-level system, we can say that the "static-field approximation" is justified for $|c_0 e \tilde{A}_0| \gg \hbar\omega_0$ and that Rabi oscillations [115] can even occur for $2m_0 c_0^2 \gg \hbar\omega_0$ if $|c_0 e \tilde{A}_0| \approx 2m_0 c_0^2$ or $|c_0 e \tilde{A}_0| \gg 2m_0 c_0^2$. With $\tilde{A}_0 = -\tilde{E}_0/\omega_0$ from $E = -\dot{A} - \nabla\phi$ with $E(t) = \tilde{E}_0 \cos(\omega_0 t)$ in a spatial antinode, this can equivalently be expressed by saying that the *"static-field approximation"* is justified for $|\xi| \gg 1$ with the dimensionless parameter

$$\xi = \frac{c_0 e \tilde{A}_0}{\hbar\omega_0} = \frac{-c_0 e \tilde{E}_0}{\hbar\omega_0^2}. \tag{4.75}$$

If, for example, the carrier photon energy is $\hbar\omega_0 = 1.5\,eV$, $|\xi| = 1 \Leftrightarrow \tilde{E}_0 = 1 \times 10^7$ V/m, equivalent to a laser intensity of $I = 2 \times 10^7$ W/cm² – a rather low value. This means that the "static-field approximation" is usually fulfilled. Furthermore, *Rabi oscillations* are expected to occur for $|\mathcal{E}| \approx 2$ or $|\mathcal{E}| \gg 2$ with

$$\mathcal{E}^2 = 4 \frac{\langle E_{kin} \rangle}{m_0 c_0^2}. \tag{4.76}$$

This form can easily be verified by inserting $\mathcal{E}$ from (4.44) and the ponderomotive energy $\langle E_{kin} \rangle$ from (4.6). *Thus, we can equivalently say that Rabi oscillations of the Dirac sea are expected if the (nonrelativistic) ponderomotive energy becomes comparable to the electron rest energy.* $|\mathcal{E}| = 10 \gg 2$ corresponds to the rather large laser intensity of $I = 1.9 \times 10^{20}$ W/cm² (see Example 4.4).

Note that such excitation of electron–positron pairs in vacuum would correspond to *nonlinear optics in vacuum*, i.e., such effects would violate the superposition principle following from the Maxwell equations in vacuum. Corresponding effects have not yet been observed in experiments. If one were to find optical nonlinearities of the vacuum, a sensitive question is whether one is able to make sure that they do not originate from residual gas atoms in the (ultrahigh) vacuum. As their nonlinear optical susceptibilities are many orders of magnitude larger than those of the vacuum, even the signal from minute atom densities might still overwhelm that of the vacuum. Direct detection of the created electron–positron pairs after a short optical pulse is not easy either because the Rabi oscillation at, e.g., $\mathcal{E} = 2$, has a period a million times $(2m_0 c_0^2/(\hbar\omega_0))$ shorter than the period of light, i.e., of order 10^{-21} s. As a result, the probability to find an electron–positron pair after the pulse is ridiculously low. This situation can be compared to that of looking for a CEO-phase dependence of the inversion w of a two-level system after an optical pulse containing 10^6 optical cycles (see Sect. 3.5).

Things look better for shorter periods of light (larger photon energies). Indeed, electron–positron pair production by inelastic light-by-light scattering of GeV photons with laser photons ($\hbar\omega_0 = 2.35\,eV$ and $I = 1.3 \times 10^{18}\,W/cm^2$) has been observed experimentally in the laboratory [116].

Real electron–positron pairs

Electron–positron pairs can also be generated in the laser field of a single propagating plane wave at yet larger intensities, as pointed out by Schwinger in 1951 [117–119]. Remember that vacuum fluctuations of the radiation field constantly generate virtual electron–positron pairs. We have discussed above that energy and momentum conservation cannot be fulfilled simultaneously in such a process. However, energy conservation can be violated for a short time span Δt determined by "time–energy uncertainty". Suppose that a photon from a vacuum fluctuation generates an electron with energy $2m_0c_0^2$. "Time–energy uncertainty" leads to $\Delta t\, 2m_0c_0^2 \approx \hbar$. Due to momentum conservation, the electron is recoiled in one direction with a velocity v near the speed of light, i.e., $v \approx c_0$. In the time interval Δt, the electron thus moves by a distance $\Delta x = v\Delta t = \lambda_c/(4\pi)$. Here we have introduced the electron *Compton wavelength*

$$\lambda_c = \frac{2\pi\hbar}{m_0c_0} = 2.4262 \times 10^{-12}\,m. \tag{4.77}$$

After time Δt, the virtual electron–positron pair again annihilates and the photon is re-emitted. If, within that time span Δt, the electron is accelerated so much by the laser field that its energy is on the order of the rest energy $m_0c_0^2$, equivalent to saying that the potential (energy) drop over length Δx, i.e., $-e\tilde{E}_0\,\Delta x$, becomes of order $m_0c_0^2$ (the other $m_0c_0^2$ comes from the positron, which is accelerated into the opposite direction), the virtual electron becomes a real electron and a real electron–positron pair has been created. In contrast to the case of counterpropagating laser beams, this pair would remain after an optical pulse. Such a process approximately happens at the *Schwinger field* $\tilde{E}_0$ given by the relation

$$\left| \frac{-e\tilde{E}_0\lambda_c}{m_0c_0^2\,4\pi} \right| = 1. \tag{4.78}$$

Note that $\tilde{E}_0$ is solely determined by the fundamental constants m_0, e, c_0 and $\hbar$. With (2.16), the resulting Schwinger field of $\tilde{E}_0 = 2.6 \times 10^{18}\,V/m$ translates into a *Schwinger intensity* of $I = 1 \times 10^{30}\,W/cm^2$. It has been estimated [120] that a single 10-fs pulse focused to a focal volume of $(1\,\mu m)^3$ with that peak intensity would generate about 10^{24} electron–positron pairs. This number reduces to one pair at $I = 1 \times 10^{27}\,W/cm^2$.

Problem 4.6. Show that we can equivalently say that the laser cyclotron energy $\hbar\omega_c$ equals the rest energy for the Schwinger field $\tilde{E}_0$, i.e.,

$$\frac{\hbar\omega_c}{2\,m_0 c_0^2} = 1\,. \tag{4.79}$$

This viewpoint emphasizes the magnetic rather than the electric-field component of the laser light.

4.6 Unruh Radiation

Zetta[2]- and Exawatt lasers with focused intensities around $I = 10^{26}$ to 10^{28} W/cm^2 might become accessible around the year 2010 and beyond. At such intensities, the peak electron acceleration a_e^0 becomes truly colossal (also see Example 4.1). For example at an intensity of $I = 10^{28}$ W/cm$^2 \Leftrightarrow \tilde{E}_0 = 2.7 \times 10^{17}$ V/m and $\tilde{B}_0 = 1.7 \times 10^6$ T in vacuum, and describing the electron in its instantaneous frame, one gets an acceleration of $|a_e^0| = e/m_0\,\tilde{E}_0 = 4.7 \times 10^{28}$ m/s$^2 = 4.8 \times 10^{27} \times$ g. This acceleration would be comparable to the gravitational acceleration near the edge of a black hole. In the latter case, the large gravitational acceleration is the origin of the so-called Hawking radiation [121], the theoretically predicted energy-loss channel of a black hole. The *Unruh radiation* [122–124] would be the analogue of that for acceleration by the laser electric field [120]. If the acceleration a_e was constant, thermal radiation according to Planck's law would result, with a temperature T given by the relation

$$k_B T = \frac{\hbar}{2\pi}\frac{a_e}{c}\,. \tag{4.80}$$

For example, for $a_e = 10^{28}$ m/s^2, $c = c_0$ and Boltzmann's constant $k_B = 1.3804 \times 10^{-23}$ J/K, one arrives at a temperature of $T = 42 \times 10^6$ K. Obviously, this Unruh radiation has to be distinguished from Bremsstrahlung, as usual originating from accelerated charges in the Maxwell equations (2.4).

---

[2] One Zettawatt=1 ZW=10^{21} W, one Exawatt=1 EW=10^{18} W.

5

Lorentz Becomes Drude: Bound–Unbound Transitions

In the previous chapters we have discussed the interaction of intense laser fields with bound electrons, promoting them from one bound state into another bound state (two-level system Bloch equations). Furthermore, we have treated free, i.e., unbound, electrons interacting with intense light fields (e.g., Volkov states). In this section, we come to the mixed case in which bound electrons are promoted into a continuum of unbound states by the light. The concepts developed for the field ionization of atoms are then further applied to other systems, such as photonuclear fission or photoemission from metal surfaces.

Before we get into all of that, we first give some simple motivation along the lines of the phenomenological approach introduced in Sect. 2.4. This already brings us to attosecond pulse trains or single attosecond pulses via high-harmonic generation from atoms. It turns out that the carrier-envelope offset phase ϕ again plays an important role.

5.1 High-Harmonic Generation: Phenomenological Approach

Macroscopically, any gas has inversion symmetry and only odd harmonics N can occur (see Sect. 2.4). What happens if, e.g., the 21st, 23rd, 25th, ... harmonic, which are evenly separated by twice the laser carrier frequency ω_0, interfer? In analogy to what we have said about mode-locking in Sect. 2.3, we anticipate a periodic train of pulses with a repetition period π/ω_0. The width of an individual pulse in that train is inversely proportional to the number of high harmonics interferring. This scenario, which has first been predicted in Refs. [125, 126], is illustrated in Fig. 5.1. Here, the duration of the individual pulses is about 350 attoseconds (one attosecond $= 1$ as $= 10^{-18}$ s). We will discuss corresponding experiments in Sect. 8.1.

In Sect. 2.6 we have seen that the interference of the fundamental with the second harmonic or the interference of the fundamental with the third harmonic can lead to a dependence on the CEO phase ϕ. In analogy to this, we also expect an influence of the CEO phase on the interference of, e.g., the ..., 79st, 81rd, 83th, ... harmonic and thus also on the shape of the train of attosecond pulses. Mathematically, we can

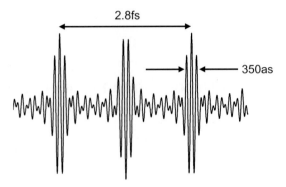

Fig. 5.1. Scheme of the electric field versus time of a train of attosecond pulses. It arises from the superposition of the 21st, 23rd, 25th, ... 31st harmonic of the fundamental wave with carrier photon energy $\hbar\omega_0 = 1.5\,eV$, equivalent to a fundamental light period of $2\pi/\omega_0 = 2.8\,fs$. All harmonics are assumed to have equal amplitude and phase, the envelope of the fundamental is taken as constant in time. Note that the period of the electric field of the pulse train is 2.8 fs, whereas that of the intensity would be 1.4 fs.

closely follow along the lines of Sect. 2.6. This leads to the general form for the high-harmonic intensity spectrum (compare with $I_{\omega_0,2\omega_0}(\omega)$ in (2.44))

$$I_{\omega_0,\,3\omega_0,\,...,79\omega_0,\,81\omega_0,\,83\omega_0,\,...}(\omega) \propto \left| \sum_{N,\,\text{odd}} e^{-iN\phi}\, E_{N\omega_0}(\omega) \right|^2 . \tag{5.1}$$

$E_{N\omega_0}(\omega)$ is the Fourier transform of the N-th harmonic with carrier frequency $N\omega_0$ resulting from a single laser pulse (rather than from a pulse train in Sect. 2.6). It depends on the details, e.g., on the electron dynamics. In order to get a feeling for the overall qualitative behavior, let us consider the simplest possible case and assume an instantaneous response according to the nonlinear optical susceptibilities in (2.37), such that the polarization $P(t)$ is a sum over terms $\propto E^N(t)$ with $E(t) = \tilde{E}(t)\cos(\omega_0 t + \phi)$. For example, for a Gaussian envelope with $\tilde{E}(t) = \tilde{E}_0 \exp(-(t/t_0)^2)$, the positive-frequency part of the Fourier transform of any power N of the envelope is again a Gaussian, i.e.

$$E_{N\omega_0}(\omega) = \omega^2 \, \eta_N \, e^{-\left(\dfrac{\omega - N\omega_0}{\sigma_{N\omega_0}}\right)^2} . \tag{5.2}$$

(Remember that Problem 2.3 has shown that the N-th harmonic of any "well-behaved" pulse shape becomes a Gaussian for $N \gg 1$.) The coefficients $\sigma_{N\omega_0}$ in the denominator of the exponent are given by

$$\sigma_{N\omega_0} = 2\sqrt{N}/t_0 . \tag{5.3}$$

Note that the spectral width is proportional to $\sigma_{N\omega_0}$ and scales as $\propto \sqrt{N}$, which substantially increases the spectral overlap of adjacent high harmonics (see Sect. 2.4).

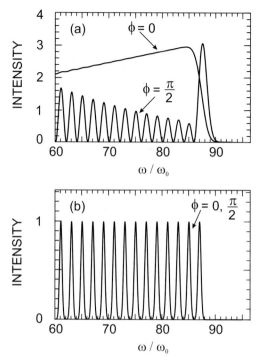

Fig. 5.2. Scheme of the intensity spectrum of high harmonics and its dependence on CEO phase ϕ according to (5.1) and (5.2). Here we have set $N^2 \eta_N = $ const. for $N = 1, 3, 5, ..., 87 = N_{\text{cutoff}}$ and 0 else, $\hbar \omega_0 = 1.5 \, eV$. The pulse duration of the incident Gaussian pulses is $t_{\text{FWHM}} = t_0 \, 2\sqrt{\ln\sqrt{2}}$. **(a)** $t_{\text{FWHM}} = 5 \, \text{fs}$, **(b)** $t_{\text{FWHM}} = 20 \, \text{fs}$. The two curves in each figure correspond to $\phi = 0, \pi, 2\pi, ...$ and $\phi = \pi/2, 3\pi/2, ...$, respectively. The latter has been stretched vertically by a factor of 4 for clarity in (a). Note that the various peaks at odd harmonics in (a) merge for "cos" 5-fs pulses, whereas the individual peaks are clearly separated for "sin" 5-fs pulses. No influence of the CEO phase is visible on this scale for 20-fs pulses, see (b). Compare with Fig. 5.6 [267].

Our reasoning implies an optically thin medium, the ω^2 factor stems from the Fourier transform of the second temporal derivative on the RHS of (2.10). The prefactors η_N depend on intensity and can be calculated in principle. The resulting dependence on the CEO phase ϕ is illustrated in Fig. 5.2(a) for 5-fs excitation pulses. For CEO phase $\phi = 0, \pi, 2\pi, ...$, the tails of the different odd harmonics add up constructively, leading to a smooth total spectrum. The Fourier transform of this smooth spectrum is a single attosecond pulse in the center of the optical pulse. For CEO phase $\phi = \pi/2, 3\pi/2, ...$, the tails of two adjacent odd harmonics interfere destructively, leading to deep valleys in between them in the spectrum. The corresponding time-domain behavior is a train of attosecond pulses. For 20-fs pulses (see Fig. 5.2(b)), the latter is true for any value of the CEO phase, because here the spectral tails of the different high harmonics hardly interfere at all (also see Fig. 5.1).

Another way to obtain attosecond pulses is to consider just *one very high harmonic* as, e.g., the $N = 101$st harmonic of a 5-fs Gaussian optical pulse with $\hbar\omega_0 = 1.5\,eV$. We have seen that the spectral width of a Gaussian to the power of N scales as $\propto \sqrt{N}$, hence, the temporal width scales as $\propto 1/\sqrt{N}$. Thus, the temporal width of the $N = 101$st harmonic is reduced by a factor of $\sqrt{101} \approx 10$, leading to a $5\,fs/10 = 500\,as$ short pulse with a carrier photon energy of $N\,\hbar\omega_0 = 151.5\,eV$. Experiments showing such *single X-ray pulses* will be discussed in Sect. 8.1.

This altogether shall be sufficient motivation to have a closer look at the microscopic physics behind high-harmonic generation from atoms.

5.2 The Keldysh Parameter

We will see that the light intensities necessary to rapidly ionize an atom are on the order of $I \approx 10^{14}$–$10^{16}\,W/cm^2$, which is well within the nonrelativistic regime (we have seen in Sect. 4.4 that the relativistic regime starts around $10^{18}\,W/cm^2$). Thus, we can ignore the laser magnetic field for the moment and focus on the laser electric field. If it were a static electric field, one could simply apply the usual rules of quantum-mechanical tunneling through potential barriers. Figure 5.3 visualizes this situation for an electron bound in the Coulomb potential of a nucleus (assumed to be much more massive or fixed in space). Here we have already tacitly used the electric-field gauge (see Sect. 4.2).

Can we really use the concept of *electrostatic* tunneling for *light* fields oscillating with a few femtoseconds period? It depends. Let us look at the problem semiclassically and call the time the electron spends *within* the barrier the electron tunneling time t_{tun}. The inverse of this time shall be the tunneling "frequency" Ω_{tun}. *This frequency must not be confused with the tunneling (or ionization) rate, related to the tunneling probability*[1]. If the tunneling time is shorter than the period of light, the laser electric field can indeed be viewed as a static field along x that parametrically changes its instantaneous value. Let us estimate the tunneling time, which is given by the width of the barrier divided by the electron velocity in the barrier: The total potential energy is $V(x) = U(x) + x\,eE(t)$, where $U(x)$ is the (Coulomb) binding potential. The latter is approximated by a rectangular potential well with finite walls (see Fig. 5.4) – a rather crude approximation (see Problem 5.1). The *width of the potential barrier, l*, depends on the instantaneous value of the electric field $E(t)$. The potential drop over length l is identical to the electron binding energy E_b. At the peak of the field, where $E(t) = \tilde{E}_0$, this leads to $l\,e\tilde{E}_0 = E_b$, equivalent to

$$l = \frac{E_b}{e\,\tilde{E}_0}\,. \tag{5.4}$$

From energy conservation, i.e., $\frac{m_e}{2}v^2 + V(x) = E_e$, we see that the electron velocity v is purely imaginary within the barrier, where the potential energy $V(x) > E_e$.

---

[1] Actually, the notion tunneling "frequency" is irritating because nothing oscillates here. One should rather call it tunneling rate – but this notion is reserved already.

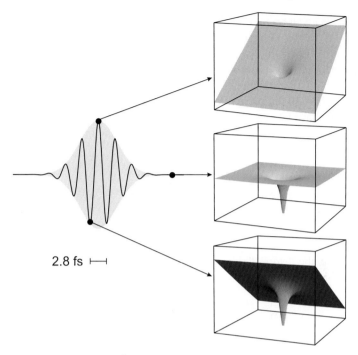

2.8 fs ⊢——⊣

Fig. 5.3. Electric field $E(t) = \tilde{E}(t)\cos(\omega_0 t + \phi)$ of a Gaussian $t_{\text{FWHM}} = 5\,\text{fs}$ linearly polarized laser pulse with carrier photon energy $\hbar\omega_0 = 1.5\,\text{eV}$ and $\phi = 0$ versus time and (two-dimensional) scheme of the resulting electric potential experienced by an electron initially bound in an atom at three characteristic points in time. The large "tilt" along the electric-field vector axis in the center of the pulse can lead to tunneling of the electron out of its binding potential through the potential barrier. If the barrier height is lowered below the binding energy, above-barrier ionization can occur. For circularly polarized light, the "tilt" stays constant but its axis rotates in time [267].

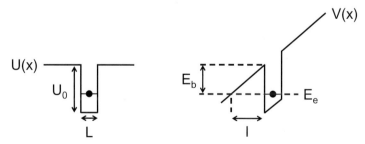

Fig. 5.4. Illustration of tunneling of an electron out of a potential well $U(x)$ (LHS) of width L and depth U_0 subject to an instantaneous electric field $E(t) > 0$ leading to a total potential $V(x)$ (RHS). The resulting potential barrier width l and the binding energy (or ionization potential) E_b are indicated. Note that in Chap. 3 on two-level systems, we have discussed excitations *within* such a potential well.

Its modulus is given by $|v(x)| = \sqrt{2(V(x) - E_e)/m_e}$. Hence, the electron velocity $|v|$ within the barrier is large if the potential barrier is high. At the same time, the tunneling rate is low. Intuitively: "The more the electron violates energy conservation within the barrier, the shorter it wants to be there." At the maximum of the barrier, where $(V(x) - E_e) = E_b$, the electron velocity is given by $|v| = \sqrt{2E_b/m_e}$. While propagating through the barrier, the electron slows down and when it has traversed the barrier, we have $V(x) = E_e$ and its kinetic energy (and the velocity) is zero. Thus, the *average electron velocity in the barrier* is roughly the mean, i.e.,

$$\langle |v| \rangle = \frac{1}{2}\left(\sqrt{2E_b/m_e} + 0\right) = \sqrt{E_b/(2m_e)}. \tag{5.5}$$

We obtain the *tunneling time*

$$t_{\text{tun}} = \frac{l}{\langle |v| \rangle} = \frac{\sqrt{2m_e E_b}}{e\tilde{E}_0}. \tag{5.6}$$

To summarize, the *"static-field approximation"* is strictly justified for

$$\frac{\Omega_{\text{tun}}}{\omega_0} \gg 1, \qquad \text{analogous to} \qquad \frac{\Omega_R}{\omega_0} \gg 1, \tag{5.7}$$

with the *peak tunneling "frequency"* $\Omega_{\text{tun}} = 1/t_{\text{tun}}$ given by

$$\Omega_{\text{tun}} = \frac{e}{\sqrt{2m_e E_b}}\tilde{E}_0, \qquad \text{analogous to} \qquad \Omega_R = \frac{d}{\hbar}\tilde{E}_0. \tag{5.8}$$

On the RHS of the last two equations, we have repeated the results of the two-level system model from Sect. 3.2 with the peak Rabi frequency Ω_R. The analogy between tunneling "frequency" and Rabi frequency is more than obvious. Both are proportional to the laser electric field and both have to be large as compared to the carrier frequency of light in order to reach the electrostatic regime[2].

The dimensionless ratio

$$\gamma_K = \frac{\omega_0}{\Omega_{\text{tun}}} = \frac{\omega_0\sqrt{2m_e E_b}}{e\tilde{E}_0} = \sqrt{\frac{E_b}{2\langle E_{\text{kin}}\rangle}} \tag{5.9}$$

is the famous *Keldysh parameter* γ_K [127] introduced by L. V. Keldysh in 1965. For $\gamma_K \ll 1$, the picture of electrostatic tunneling applies. On the RHS of (5.9), we have alternatively expressed the Keldysh parameter using the ponderomotive energy $\langle E_{\text{kin}}\rangle$, which can easily be verified by insertion of (4.6) into (5.9). Thus, we can equivalently say that *something special is expected to happen if the peak kinetic energy* $2\langle E_{\text{kin}}\rangle$, *added to the electron by the laser electric field, becomes comparable to the electron binding energy* E_b.

---

[2] We will see in Chap. 7 that Ω_R is indeed nearly identical to Ω_{tun} for semiconductors.

▶ **Example 5.1.** For an electron in the 1s state of a hydrogen atom, E_b is identical to the Rydberg energy of 13.6 eV (also see Table 5.1), $m_e = m_0$. With a carrier photon energy of $\hbar\omega_0 = 1.5$ eV, equivalent to $2\pi/\omega_0 = 2.8$ fs period of light, unity Keldysh parameter corresponds to

$$\gamma_K = 1 \tag{5.10}$$

$$\Leftrightarrow$$

$$t_{tun} = 0.44 \text{ fs}$$

$$\Leftrightarrow$$

$$\tilde{E}_0 = 2.8 \times 10^{10} \text{ V/m}$$

$$\Leftrightarrow$$

$$I = 1.0 \times 10^{14} \text{ W/cm}^2.$$

For the same parameters, we have the barrier width $l = 0.5$ nm and the average electron velocity in the barrier $\langle|v|\rangle = 1 \times 10^6$ m/s $\ll c_0 = 3 \times 10^8$ m/s. The latter inequality shows that our nonrelativistic treatment is meaningful indeed.   ◀

Table 5.1. Ionization potential E_b (in units of eV) of selected relevant gases (for single ionization).

	H	He	Ne	Ar	Kr	Xe
E_b	13.598	24.587	21.564	15.759	13.99	12.127

It is instructive to compare the peak laser electric field $\tilde{E}_0$ of Example 5.1 with the electric field attracting the electron to the nucleus. For the 1s state of a hydrogen atom, we have

$$|E|_{H, 1s} = \frac{e}{4\pi\epsilon_0 r_B^2}. \tag{5.11}$$

With the hydrogen Bohr radius of $r_B = 0.053$ nm, we obtain an electric field of

$$|E|_{H, 1s} = 5.17 \times 10^{11} \text{ V/m}. \tag{5.12}$$

Thus, the peak laser electric field matches the inner-atomic field, i.e., $\tilde{E}_0 = |E|_{H, 1s}$, for $\gamma_K = 0.05 \ll 1$, equivalent to an intensity of $I = 3.4 \times 10^{16}$ W/cm^2.

5.3 Field Ionization of Atoms

Within the "static-field approximation", the *ionization rate* $\Gamma_{ion}(t)$ (or tunneling rate) depends exponentially on the instantaneous barrier width $l(t)$ because the electron wave function is decaying exponentially in the barrier according to $\psi(x) \propto$

$\exp(-|k_x(x)|\,x)$. The probability of tunneling through the barrier is $\propto |\psi(l)|^2$. Approximating $|k_x(x)| \rightarrow \langle|k_x|\rangle$ with $\hbar\langle|k_x|\rangle = m_e\langle|v|\rangle$ using (5.5) and inserting $l(t) = E_b/(e\,|E(t)|)$ analogous to (5.4), we expect the general behavior

$$\frac{\Gamma_{ion}(t)}{\Gamma^0_{ion}} = e^{-\frac{\sqrt{2m_e E_b}}{\hbar} l(t)} = e^{-\frac{\frac{1}{\hbar e}\sqrt{2m_e}\,E_b^{3/2}}{|E(t)|}} = e^{-\frac{E_{exp}}{|E(t)|}}. \qquad (5.13)$$

This dependence is illustrated in Fig. 5.5. It exhibits a *threshold behavior*, i.e., above a certain value of the instantaneous laser electric field $|E(t)|$, the ionization rate increases steeply[3]. Based on this finding, we expect a strong dependence of the ionization on the CEO phase ϕ of the exciting laser pulses, as schematically depicted in Fig. 5.6. Note that this finding is consistent with the expectation from the above phenomenological approach (compare with Fig. 5.2). For few-cycle pulses, electrons generated by *above-threshold ionization* are primarily emitted towards one side for $\phi = 0$ and to the other side for $\phi = \pi$, while they go to both sides with equal probability for CEO phase $\phi = \pm\pi/2$. This effect, which clearly disappears for pulses containing many cycles of light, has also been observed experimentally [128]. Further related experiments with few-cycle laser pulses are discussed in Sect. 8.1.

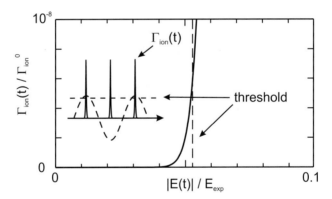

Fig. 5.5. Illustration of the instantaneous ionization rate $\Gamma_{ion}(t)$ versus the instantaneous laser electric field $|E(t)|$ in the "static-field approximation" according to (5.13). The effect of the indicated "threshold" is illustrated in Fig. 5.6. The inset illustrates the laser carrier-wave oscillation and the ionization rate $\Gamma_{ion}(t)$ versus time [267].

---

[3] Note that the electric fields with $|E(t)|/E_{exp} < 0.1$ in Fig. 5.5 are barely compatible with the electrostatic limit as $\gamma_K = \frac{\hbar\omega_0}{E_b}\frac{E_{exp}}{|E(t)|}$ is about unity for $|E(t)|/E_{exp} = 0.1$, $E_b = 13.6\,eV$, $m_e = m_0$ and $\hbar\omega_0 = 1.5\,eV$ ($E_{exp} = 2.6 \times 10^{11}$ V/m). Nevertheless, the depicted behavior is qualitatively correct. Mathematically, the two limits $|E(t)|/E_{exp} \ll 1$ and $\gamma_K \ll 1$ can strictly be fulfilled simultaneously for $E_b \gg \hbar\omega_0$.

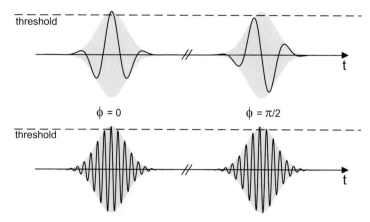

Fig. 5.6. For excitation of an atom with laser pulses $E(t) = \tilde{E}(t)\cos(\omega_0 t + \phi)$ with, e.g., Gaussian envelope $\tilde{E}(t) = \tilde{E}_0 \exp(-(t/t_0)^2)$, the actual peak electric field can be above the ionization threshold (see Fig. 5.5) for CEO phase $\phi = 0$ (LHS column), while it is below the threshold for $\phi = \pi/2$ (RHS column). This effect is expected to be pronounced for short laser pulses (upper part) and a minor detail for long pulses (lower part). The gray areas indicate the pulse envelopes. Compare with Fig. 5.2 [267].

One can still use the concept of electron tunneling for $\gamma_K \geq 1$ or even $\gamma_K \gg 1$, but there the potential changes during the tunneling process. Thus, simple tunneling formulae are no longer available. We can, however, employ numerical solutions of the time-dependent Schrödinger equation. Let us, for simplicity, again consider its one-dimensional version

$$i\hbar\frac{\partial}{\partial t}\psi(x,t) = \left(-\frac{\hbar^2}{2m_e}\frac{\partial^2}{\partial x^2} + U(x) + x\,e\,E(t)\right)\psi(x,t). \qquad (5.14)$$

with the laser electric field

$$E(t) = \tilde{E}(t)\cos(\omega_0 t + \phi) \qquad (5.15)$$

and the binding potential $U(x)$. We again choose $U(x)$ to be a simple potential well, i.e., $U(x) = -U_0$ for $|x| \leq L/2$ and $U(x) = 0$ else. The simulated region is significantly larger than the one depicted in Fig. 5.7. Starting with the ground state wave function of the finite potential well at $t = 0$ in Fig. 5.7, the real part of the electron wave function oscillates with a frequency given by $E_b/\hbar$ as expected from elementary quantum mechanics ($E_b \approx U_0$ for $L = 0.6$ nm). As the instantaneous electric field $E(t)$ increases from $E(t=0) = 0$ towards its peak $E(t=0.7\,\text{fs}) = \tilde{E}_0$ in Fig. 5.7, the bound part of the wave packet is shifted towards the left. This motion of charge corresponds to optical transitions *within the well* and has been discussed extensively on the basis of the two-level system Bloch equations in Sect. 3.2. A comparison with Fig. 3.2 shows that the wave function in the well exhibits more structure in

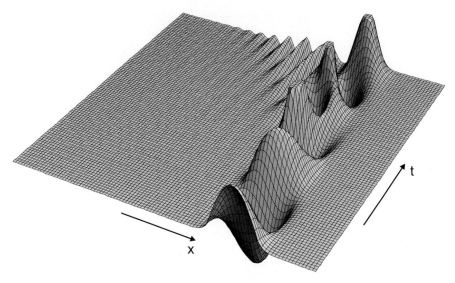

Fig. 5.7. Real part of the electron wave function, $\mathrm{Re}(\psi(x, t))$, versus x-coordinate and time t from numerical solution of the time-dependent Schrödinger equation (5.14) within the electric-field gauge. x runs from $-3\,\mathrm{nm}$ to $+1\,\mathrm{nm}$, t from 0 to 1.0 fs. The electron is initially ($t = 0$) in the ground state of a potential well $U(x)$ of depth U_0 and width L (also see Fig. 5.4). Parameters are $\phi = -\pi/2$, $\hbar\omega_0 = 1.5\,e\mathrm{V}$ equivalent to a 2.8-fs period of light, $m_e = m_0$, $L = 0.6\,\mathrm{nm}$ and $U_0 = 15\,e\mathrm{V} \approx E_b$. The constant electric field envelope of $\tilde{E}_0 = 3 \times 10^{10}\,\mathrm{V/m}$ corresponds to an intensity of $I = 1.1 \times 10^{14}\,\mathrm{W/cm^2}$ or to $\frac{\langle E_{\mathrm{kin}}\rangle}{\hbar\omega_0} = 5$ and to a Keldysh parameter of $\gamma_K \approx 1$. Compare with Fig. 4.3 for a free vacuum electron and with Fig. 3.2 where only two bound states are accounted for.

Fig. 5.7, indicating that not only transitions from the ground state to the next excited state are involved but also transitions into and/or between higher excited states of the well. In addition to this, the wave function also reveals contributions *outside the binding potential* that correspond to tunneling out of the well. These contributions are accelerated by the electric field (also compare Fig. 4.3 for free electrons) and eventually give rise to high-harmonic generation, which will be discussed below. Note the delay between the maximum of the electric field at $t = 0.7\,\mathrm{fs}$ and the emission of the electron wave packet to the left (the end of the depicted timescale is 1.0 fs). This delay is due to the tunneling time t_{tun}. Indeed, the numbers of Example 5.1 (which imply a constant electric field) are roughly consistent with the numerical simulation. Also, as expected from the above semiclassical threshold discussion, tunneling is largely suppressed if the peak electric field $\tilde{E}_0$ is reduced by just a factor of two (see Fig. 5.8).

What can be done for $\gamma_K \gg 1$ apart from "brute-force" numerical calculations ? At this point it can be advantageous to switch from the electric-field gauge to the radiation gauge (see Sect. 4.2). Indeed, it is attractive mathematically and has been used for $\gamma_K \gg 1$ [129–131]. This brings us into the regime of *multiphoton absorption*. The laser electric field is not too strong and we are interested in transitions from the bound

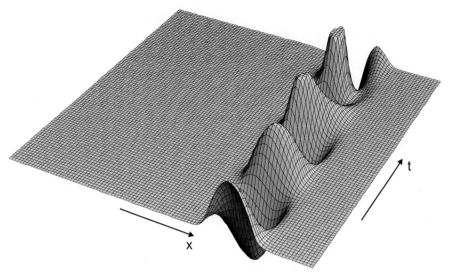

Fig. 5.8. As Fig. 5.7, but for $\tilde{E}_0 \rightarrow \tilde{E}_0/2$, thus $\gamma_K \approx 2$. Note that tunneling out of the well is significantly reduced with respect to Fig. 5.7.

state into the unbound states. The influence of the laser field on the bound state can be ignored to a first approximation. Furthermore, the unbound states become identical to Volkov states if we neglect the influence of the Coulomb binding potential on them. As discussed in Sect. 4.2, the Volkov states consist of a series of N-photon sidebands, which gain weight as the laser intensity increases. Remember that, within the Volkov states, the influence of the laser electric field is incorporated nonperturbatively. The *two regimes* of *electrostatic tunneling* ($\gamma_K \ll 1$) on the one hand and *multiphoton absorption* ($\gamma_K \gg 1$) on the other hand are illustrated in Figs. 5.9 (a) and (b), respectively.

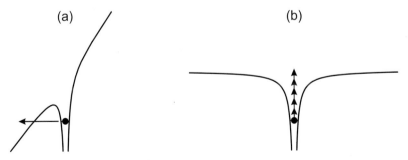

Fig. 5.9. Illustration of ionization of atoms via (a) electrostatic tunneling at large laser fields (Keldysh parameter $\gamma_K \ll 1$) and (b) multiphoton absorption at weak laser electric fields ($\gamma_K \gg 1$). According to Example 5.1, $\gamma_K = 1$ corresponds to a laser intensity of 10^{14} W/cm^2 for typical parameters [267].

It is interesting to compare the numbers from Example 5.1 on the electrostatic limit with those from Example 4.2 on Volkov states. In the latter example, we have estimated that the amplitude of the N-photon sideband of a free vacuum electron has a maximum around an intensity of $I = N \times 3 \times 10^{13}$ W/cm^2 for $N \gg 1$. For the parameters of Example 5.1, i.e., $E_b = 13.6\,eV$ and $\hbar\omega_0 = 1.5\,eV$, the integer N would have to be $E_b/(\hbar\omega_0) = 13.6/1.5 > 9$ to make a real transition from the bound 1s state into the ionization continuum. $N = 10$ yields an intensity of $I = 4 \times 10^{14}$ W/cm^2, which is amazingly close to $I = 1 \times 10^{14}$ W/cm^2 corresponding to $\gamma_K = 1$ from Example 5.1: Two rather different estimates and physical pictures give consistent results.

▶ **Example 5.2.** The interaction of intense light with nuclei [132] can either be direct or can be mediated by atomic electrons that are accelerated to MeV ponderomotive energies by the laser field (see Sects. 4.4 and 4.5). Here, we only consider the first case because it is similar to the photoionization of atoms. By analogy, we anticipate photonuclear "ionization", i.e., fission. In contrast to the atomic Coulomb potential, however, the potential U an α-particle experiences in the nucleus is composed of the binding part arising from the strong interaction and a repulsive part due to the Coulomb repulsion of the positively charged α-particles. As a result, even without a laser field, the α-particle can tunnel out of the nucleus – the normal radioactive α-decay. For large laser electric fields, we can ignore these "details" to lowest order and use the Keldysh parameter from (5.9), replacing $m_e \rightarrow m_\alpha$, with the α-particle mass $m_\alpha = 6.7 \times 10^{-27}$ kg to estimate the relevant fields and intensities. For $\hbar\omega_0 = 1.5\,eV$ and a typical binding energy of $E_b = 5\,MeV$ (still, $v \ll c_0$ according to (5.5)), unity Keldysh parameter corresponds to a peak of the laser electric field of $\tilde{E}_0 = 1 \times 10^{15}$ V/m, equivalent to a peak intensity of $I = 2 \times 10^{23}$ W/cm^2. Thus, the electrostatic regime with $\gamma_K \ll 1$ cannot be reached with current lasers. The experimentally observed photonuclear fission at intensities ranging from $I = 10^{19}$ W/cm^2 [133] to $I > 10^{20}$ W/cm^2 [134], pulse durations of a few 100 fs and $\hbar\omega_0 = 1.2\,eV$ is rather mediated by atomic electrons in solid targets.   ◀

Problem 5.1. In our discussion of electrostatic tunneling we have approximated the binding potential as a rectangular potential well. For the actual Coulomb potential, above-barrier ionization can take place (see Fig. 5.3). Estimate the laser electric field at which this process appears. This field is called the *barrier-suppression field* [135]. What are the consequences for the ionization rate?

5.4 High-Harmonic Generation

Having solved the time-dependent Schrödinger equation for the ionization numerically in the previous section, we can *in principle* simply go ahead and compute the atomic dipole moment from the known instantaneous wave function $\psi(r, t)$ via the expectation value $\langle \psi(r, t)| - e\, r\, |\psi(r, t)\rangle$. Multiplying by the density of atoms delivers the macroscopic optical polarization P, which enters into the Maxwell equations.

Neglecting propagation effects, the radiated electric field is proportional to the second temporal derivative of P. The square modulus of its Fourier transform delivers the high-harmonic (intensity) spectrum. This can and has been worked out [136–138]. Indeed, Fig. 5.2(b) can be viewed as a scheme of the resulting high-harmonic spectrum. Typically, the intensity of the harmonic orders decays rapidly over several orders of magnitude up to about order ten or twenty (not shown). This is followed by a *plateau region* of more or less constant harmonic orders and a rapid fall for orders above the *cutoff order* N_{cutoff}. The latter depends on the laser intensity and the atom or ion under consideration.

5.4.1 Three-Step Scenario and Cutoff

It is, however, instructive and much more intuitive to discuss the problem semiclassically in terms of the so-called *three-step scenario* introduced by Corkum [139–142]. It turns out that the results are qualitatively identical and, moreover, that they do not even depend much on the particular choice of the binding potential $U(r)$.

Semiclassically, the atom is ionized at some instant in time (step # 1). According to (5.13) and/or Fig. 5.5, the instantaneous ionization rate is going to peak when the modulus of the laser electric field has a maximum. Let us consider *linear polarization* of the laser. Starting with zero velocity and potential energy E_b, the electron is accelerated (step # 2) in the laser electric field and again decelerated when the field changes sign in the following half-cycle (see Fig. 4.3). After a few femtoseconds it can come back to its location of birth for the first time. At this point, its total energy E_e is the sum of the binding energy E_b and the kinetic energy acquired since its birth. Classically, it will pass the nucleus. Quantum mechanically, it can emit a photon, which takes all the electron energy, and fall back into the bound state (step # 3). The maximum photon energy $\hbar\omega = N_{cutoff}\,\hbar\omega_0 = E_e$ to be expected is thus directly related to the maximum electron energy E_e. Once the atom is ionized, the Coulomb field of the nucleus can be neglected and the electron essentially acts like a free electron. Following Sect. 4.1 on free electrons, we simply have to solve Newton's law, i.e., $m_e\ddot{x} = -eE(t)$, with $E(t) = \tilde{E}(t)\cos(\omega_0 t + \phi)$ with $\phi = 0$ for the initial conditions that at the time of birth t_0, the electron coordinate and velocity are zero, i.e., $x(t_0) = 0$ and $v(t_0) = 0$. For a constant field envelope $\tilde{E}(t) = \tilde{E}_0$, the solution is

$$x(t) = \frac{e\tilde{E}_0}{m_e\omega_0^2}\left[\left(\cos(\omega_0 t) - \cos(\omega_0 t_0)\right) + \sin(\omega_0 t_0)(\omega_0 t - \omega_0 t_0)\right] \qquad (5.16)$$

and

$$v(t) = -\frac{e\tilde{E}_0}{m_e\omega_0}\left(\sin(\omega_0 t) - \sin(\omega_0 t_0)\right), \qquad (5.17)$$

resulting in the electron kinetic energy at time t

$$\frac{m_e}{2}v^2(t) = 2\langle E_{kin}\rangle\left(\sin(\omega_0 t) - \sin(\omega_0 t_0)\right)^2. \qquad (5.18)$$

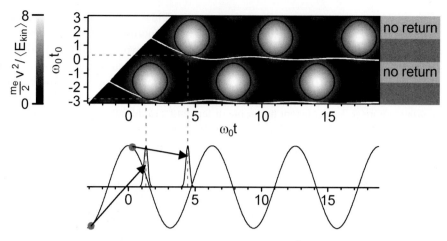

Fig. 5.10. Grayscale image of the electron kinetic energy $\frac{m_e}{2} v^2$ in units of the ponderomotive energy $\langle E_{kin} \rangle$ according to (5.18) versus time t and time of ionization t_0 for an electric field $E(t) = \tilde{E}_0 \cos(\omega_0 t)$ (see bottom). The unphysical range of $t < t_0$ is hidden. The isoenergy lines $\frac{m_e}{2} v^2 / \langle E_{kin} \rangle = 3.17$ are shown by the six closed black curves. The condition $x = 0$ (position of the nucleus) from (5.16) is depicted by the two white solid curves. The two equivalent points of maximum electron kinetic energy, $3.17 \langle E_{kin} \rangle$, at the electrons first return to $x = 0$ are shown by the dashed lines. On the RHS, regions of birth phases $\omega_0 t_0$ with no return to the nucleus at all are indicated. The lower part shows the laser electric field with those birth phases $\omega_0 t_0$ giving maximum kinetic energy as full gray dots. At the corresponding recollision phases $\omega_0 t$, EUV pulses (schematically) are emitted. This timing has been confirmed experimentally [143]. Note that the different electron energies correspond to different harmonics and to slightly different recollision phases. This introduces a certain chirp on the EUV pulse [268].

Here we have introduced the ponderomotive energy $\langle E_{kin} \rangle$ according to (4.6). For example, for $\omega_0 t_0 = 0$, we recover (4.5) for $\phi = 0$ and the peak electron kinetic energy is $2\langle E_{kin} \rangle$, for $\omega_0 t_0 = \frac{\pi}{2}$, the peak is $8\langle E_{kin} \rangle$. We, however, want to compute the maximum kinetic energy at the time $t = t_1 > t_0$, when the electron comes back to the nucleus at $x = 0 = x(t_1)$ for the first time. Because of the 2π periodicity, it is sufficient to consider the interval $[-\pi, +\pi]$ of birth phases $\omega_0 t_0$. According to (5.16), for $\omega_0 t_0$ in the interval $[-\pi, -\frac{\pi}{2}]$, the electron comes back to $x = 0$, for $[-\frac{\pi}{2}, 0]$ it never comes back, for $[0, \frac{\pi}{2}]$ it does come back, for $[\frac{\pi}{2}, \pi]$ it does not. Numerical or graphical (see Fig. 5.10) solutions of (5.16) and (5.18) show that the maximum electron kinetic energy at its first return to $x = 0$ occurs at time $\omega_0 t = \omega_0 t_1 \approx 1.3$ for the birth phase $\omega_0 t_0 \approx -\pi + 0.3$. An equivalent maximum occurs at time $\omega_0 t_1 \approx 1.3 + \pi$ for $\omega_0 t_0 \approx +0.3$. Inserting these numbers into (5.18), we obtain a kinetic energy of $3.17 \langle E_{kin} \rangle$, the total electron energy at this point is $E_e = E_b + 3.17 \langle E_{kin} \rangle$ (see Fig. 5.11). At these birth phases, the electric field is $|E(t_0)| = \tilde{E}_0 |\cos(\omega_0 t_0)| = \tilde{E}_0 \times 0.96$, thus the instantaneous ionization rate according to (5.13) is rather high. For example, for $\omega_0 t_0 = 0$, where the ionization rate has its absolute peak, the electron returns at $\omega_0 t = \omega_0 t_1 = 2\pi$, leading to zero

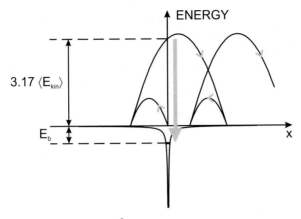

Fig. 5.11. Electron kinetic energy $m_e v^2(t)/2$ versus its coordinate $x(t)$ according to (5.18) and (5.16) starting at the time of ionization t_0 at $x = 0$. For the depicted birth phase of $\omega_0 t_0 = 0.3$, the kinetic energy at the first (in this case the only) return of the electron to the nucleus at $x = 0$ has the maximum of $3.17 \langle E_{\text{kin}} \rangle$. The resulting emitted harmonic photon energy $\hbar\omega = N_{\text{cutoff}} \hbar\omega_0 = E_b + 3.17 \langle E_{\text{kin}} \rangle$ is indicated by the thick arrow. The lower part shows the Coulomb binding potential [267].

kinetic energy. The scenario just described is periodic in time with frequency ω_0. Thus, the highest harmonic-order, the cutoff, is given by the odd integer nearest to

$$N_{\text{cutoff}} = \frac{E_b + 3.17 \langle E_{\text{kin}} \rangle}{\hbar\omega_0}. \tag{5.19}$$

Note that the cutoff harmonic is linear in the laser intensity $I \propto \langle E_{\text{kin}} \rangle$. This is in contrast to the corresponding expression for two-level systems, (3.33), where N_{cutoff} is proportional to $\sqrt{I} \propto \Omega_R$ in the "static-field limit". Both approaches do, however, agree in that a cutoff does exist. This is not *a priori* obvious at all. One could have expected a continuous decrease of the strength of the harmonics with harmonic order N rather than a sharp cutoff. Indeed, e.g., for high-harmonic generation via relativistic nonlinear Thomson scattering on free electrons (see Sect. 4.4.2) no such cutoff appears.

▶ **Example 5.3.** For a peak laser intensity of $I = 4 \times 10^{14} \, \text{W/cm}^2$ ($\Leftrightarrow \gamma_K = 0.5$ or $\tilde{E}_0 = 5.6 \times 10^{10} \, \text{V/m}$ from Example 5.1), thus $\langle E_{\text{kin}} \rangle = 27.2 \, eV$ in vacuum, $\hbar\omega_0 = 1.5 \, eV$, the 1s state of hydrogen with $E_b = 13.6 \, eV$, and $m_e = m_0$, (5.19) results in a cutoff of $N_{\text{cutoff}} = 67$. This is equivalent to a maximum photon energy of $100.5 \, eV$ or a minimum wavelength of $12.3 \, \text{nm}$.
The corresponding electron trajectory according to (5.16) for the birth phase $\omega_0 t_0 = 0.3$ exhibits an excursion after ionization and before the first return to the nucleus of $e\tilde{E}_0/(m_0 \omega_0^2) \times 1.2 = 2.4 \, \text{nm}$. This is 46 times the hydrogen Bohr radius of $r_B = 0.053 \, \text{nm}$. ◀

If one is interested in making N_{cutoff} as large as possible, (5.19) favors large binding energies (ionization potentials, see Table 5.1) for the price of larger ionization thresholds according to (5.13). At some point, however, the electron motion becomes relativistic (as $|v| \propto \sqrt{E_b}$ according to (5.5)) and high harmonics with wavelengths less than 0.1 nm occur in numerical studies [144]. For a given type of atoms, hence fixed E_b, the cutoff obviously increases with increasing ponderomotive energy. Note, however, that we have tacitly assumed in our above semiclassical reasoning that the atom has not been ionized yet. Thus, for pulsed excitation, $\langle E_{\text{kin}} \rangle$ in (5.19) has to be interpreted as the electron ponderomotive energy in the first optical cycle after ionization. This introduces an *implicit dependence on pulse duration:* For long and weak pulses, the atom is ionized in the center of the pulse and the ponderomotive energy is small, hence the cutoff is low. For long and intense pulses the atom is ionized well before the maximum of the pulse envelope is reached (see Fig. 5.12(a)) and, again, the ponderomotive energy is low. For short *and* intense pulses, the atom is ionized near the center of the pulse where the intensity and the ponderomotive energy are large, thus, the cutoff is shifted towards high harmonics.

Mathematically, the *degree of ionization* can simply be obtained by integration of the tunneling or ionization rate (5.13), which has followed from our semiclassical, electrostatic reasoning (also see Problem 5.1). Considering single ionization only,

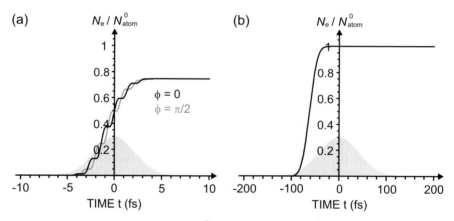

Fig. 5.12. Degree of ionization, $N_e(t)/N_{\text{atom}}^0$, according to (5.20) and (5.13), versus time t. **(a)** Gaussian optical pulses with $t_{\text{FWHM}} = 5\,\text{fs}$ and $\phi = 0$ (black curve) and $\phi = \pm\pi/2$ (gray curve), **(b)** corresponding 100-fs pulses (dependence on ϕ vanishes within the line thickness). The gray areas illustrate the pulse intensity, note the different timescales. All other parameters are identical, especially the peak intensity is the same, i.e., $\bar{E}_0/E_{\text{exp}} = 0.5$, $\hbar\omega_0 = 1.5\,\text{eV}$, and $\Gamma_{\text{ion}}^0 = 5/\text{fs}$. Nevertheless, this figure should be viewed as a schematic rather than as a quantitative calculation. Note that in (b) essentially all of the atoms are already ionized when the peak intensity and hence the peak ponderomotive energy are reached in the center of the pulse at $t = 0$. This inhibits efficient generation of high-harmonic orders for (b) – in sharp contrast to (a), even though the pulse energy is 20 times *lower* in (a). By the way: Note that the degree of ionization long after the pulse in (a) shows no significant dependence on ϕ (see Problem 5.3). It is interesting to compare the latter aspect with Figs. 3.9 and 3.10 [269].

and starting from a number of atoms N_{atom}^0 prior to excitation, the number of ionized atoms equals the number of free electrons $N_e(t) \leq N_{atom}^0$, thus the remaining number of atoms that are *not* ionized is given by $N_{atom}(t) = N_{atom}^0 - N_e(t)$. Together with the instantaneous ionization rate $\Gamma_{ion}(t) \geq 0$ according to (5.13), we have

$$\frac{dN_e}{dt} = \Gamma_{ion}(t)\, N_{atom}(t) . \tag{5.20}$$

The formal solution is

$$N_e(t) = N_{atom}^0 \left[1 - \exp\left(-\int_{-\infty}^{t} \Gamma_{ion}(t')\, dt' \right) \right] . \tag{5.21}$$

The degree of ionization, a number that monotonously increases from zero to some value ≤ 1, is given by $N_e(t)/N_{atom}^0$. The numerical solutions shown in Fig. 5.12 confirm our above qualitative reasoning. Note that the electrostatic approximation underlying (5.13) becomes questionable in the wings of the pulses, where the electric field is low, hence the Keldysh parameter is large. Indeed, numerical solutions of the time-dependent Schrödinger equation (not shown) tend to exhibit a yet more drastic dependence on the pulse duration.

Problem 5.2. In the semiclassical treatment of ionization and high-harmonic generation, we have considered linear polarization of light. What changes for *circular polarization*?

Problem 5.3. Notably, the degree of ionization for $t \rightarrow \infty$ in Fig. 5.12(a) exhibits *no* significant dependence on the CEO phase ϕ. This appears to contradict our expectation from Fig. 5.6 together with Fig. 5.5 for few-cycle pulses. Can you resolve this riddle?

Problem 5.4. Discuss the physics described in Figs. 5.7 and 5.8 with $\gamma_K \approx 1$ and $\gamma_K \approx 2$, respectively, in terms of the validity of the two-level system approximation for transitions between the ground state and the first excited state of the potential well.

Problem 5.5. Compare Figs. 3.4 and 5.12. $w(t)$ in Fig. 3.4 addresses the inversion of a two-level system versus time, Fig. 5.12 the corresponding degree of ionization of an atom. They both exhibit an oscillatory component with twice the carrier frequency of light and an overall increase on a longer timescale. However, there are distinct differences as well: The rapid oscillation in Fig. 3.4 can lead to an instantaneous decrease, in Fig. 5.12 we have a strictly monotonous increase. The envelope in Fig. 3.4 can decrease (the Rabi flopping), whereas it monotonously increases with time in Fig. 5.12. We have discussed the mathematics. Try to give an intuitive explanation for these qualitative differences.

5.5 Application to Photoemission from Metal Surfaces

A laser electric-field vector with a significant component parallel to the normal of a metal/vacuum interface can lead to photoemission of electrons (see Fig. 5.13). Here, the binding energy (ionization potential) of the atom, E_b, has to be replaced by the metal work function W – the difference between the vacuum level and the metal Fermi energy E_F, or, in other words, the minimum work that needs to be supplied to rip a crystal electron out of the metal. Typical electron effective masses m_e are close to the free electron mass m_0, typical values for W lie in the range of 2–5 eV (see Table 5.2). Provided that the carrier photon energy follows $\hbar\omega_0 < W$, linear photoemission cannot occur. For example, for $W = 3\,eV$, $m_e = m_0$ and $\hbar\omega_0 = 1.5\,eV$, unity Keldysh parameter γ_K corresponds to a peak laser intensity of $I = 2 \times 10^{13}\,W/cm^2$, which is just barely compatible with typical damage thresholds. Hence, the electrostatic regime ($\gamma_K \ll 1$) can hardly be reached.

Table 5.2. Work function W (in units of eV) of selected metals. Values taken from Ref. [145].

	Na	Cs	Cu	Ag	Au	Fe	Al	W
W	2.35	1.81	4.4	4.3	4.3	4.31	4.24	4.5

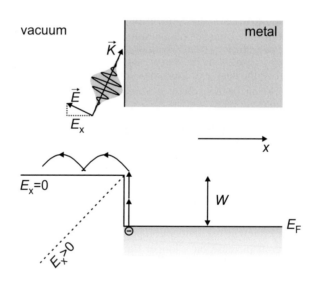

Fig. 5.13. Schematic illustration of photoemission from a metal/vacuum interface, e.g., corresponding to the cathode of a photomultiplier tube. Upper part: The linearly polarized few-cycle optical pulse with carrier photon energy $\hbar\omega_0 \gtrsim W/2$, wavevector $\mathbf{K}$ and electric-field vector $\mathbf{E}$ impinges under nearly grazing incidence. Lower part: An electron at the Fermi energy E_F of the metal experiences an energy barrier of height W – the metal work function. Inside the metal, the large electron density efficiently screens the laser electric field except for a thin surface layer. The gray area represents the Fermi sea. After ionization, the electron moves towards the LHS. Its trajectory $E_e(x)$ as obtained from (5.16) and (5.18) with $\omega_0 t_0 \to \phi$ for $\phi = -\pi/4$ and $\gamma_K \approx 3$ is depicted (compare with Fig. 5.11 for $\omega_0 t_0 = +0.3$) [269].

We can now take advantage of our discussion on electron "return" or "no return" in Fig. 5.10 in order to qualitatively discuss the dependence of the photoemission on the CEO phase ϕ within the multiphoton regime ($\gamma_K \gg 1$). In this regime, the tunneling time exceeds the period of light, hence the carrier-wave oscillation is averaged out and the photoelectron emission peaks at the maximum of the electric field *envelope* of the pulse. Let us assume that this maximum lies at time $t = 0$ and discuss an electron freed right at this time. In order to eventually contribute to the photocurrent, the freed electron has to arrive at the anode (or the first dynode) at some point, i.e., it must *not* return to the metal surface. As soon as the electron is in vacuum, it is subject to the instantaneous laser electric field $E(t)$, which depends on the CEO phase ϕ. Thus, we can reinterpret the birth phase $\omega_0 t_0$ in Fig. 5.10 by the CEO phase ϕ, i.e., replace $\omega_0 t_0 \rightarrow \phi$ (to get the correct sign, Fig. 5.6 might help). As discussed above (Sect. 5.4.1), the emitted electrons move towards the LHS and do not come back to $x = 0$ (the metal surface) for the ϕ-interval $[-\frac{\pi}{2}, 0]$ – the photocurrent is large. Note that *the metal/vacuum interface breaks the inversion symmetry* with respect to the surface normal (x-direction). Thus, the upper region of "no return" in Fig. 5.10 does not apply here, because the electric force would lead to a motion of the electrons towards the RHS (see Fig. 5.11), i.e., the electrons would be pushed into the metal. A density fluctuation in the metal results – the photocurrent is low. *Altogether, we anticipate that the time-integrated photocurrent has a component that oscillates versus ϕ with period 2π.* For a constant pulse envelope, this dependence would completely average to zero, whereas for few-cycle pulses, the temporal variation of the envelope within one optical cycle leads to a finite modulation. Theoretical calculations [146] based on the jellium model show that such ϕ-dependence does indeed occur. It turns out that for 5-fs pulses with a carrier photon energy of $\hbar\omega_0 = 1.5\,eV$ and a Keldysh parameter of $\gamma_K \approx 2-3$, the absolute photocurrent as well as the relative modulation depth have appreciable magnitude. Maximum photocurrent is predicted at $\phi = -\frac{\pi}{4}$, the middle of the interval $[-\frac{\pi}{2}, 0]$ (see trajectory in Fig. 5.13). Corresponding experiments using a commercial photomultiplier cathode (*Hamamatsu R595*) within the multiphoton regime (gold cathode, $I \approx 2 \times 10^{12}\,W/cm^2$, 4-fs pulses, $\hbar\omega_0 = 1.65\,eV$) have been reported in Refs. [147] and [148].

At the end of Sect. 2.3 we have seen that any optical pulse has to obey the condition $\int_{-\infty}^{+\infty} E(t)\,dt = 0$, i.e., the average force on the electron is zero. Consequently, the electron velocity is zero long after the pulse and the electron never actually arrives at the first dynode (which is typically rather far away) unless one applies an additional bias voltage – as is, of course, done in these experiments for the multiplication process anyway.

Alternatively, we can think about the electron current towards the LHS ($x < 0$) as transitions from electron states within the metal into those vacuum Volkov states with $k_x < 0$. The latter condition breaks inversion symmetry. Indeed, in Sect. 4.2 we have seen that the N-photon Volkov sideband (see Fig. 4.2) has a phase $N\phi$ (see (4.18)). Thus, loosely speaking, e.g., for $3\hbar\omega_0 > W > 2\hbar\omega_0$, two-photon absorption ($N = 2$) from the high-energy part of the laser spectrum can interfere with three-photon absorption ($N = 3$) from the low-energy part (somewhat analogous to our discussion in Sect. 3.5, except for the symmetry). This leads to a beating with difference phase ϕ, hence to a period of 2π. In contrast, for $k_x = 0$ and according to (4.22), only even-order Volkov sidebands would occur, hence the beating would have difference phase 2ϕ, thus the oscillation period would equal π rather than 2π. Thus, the breaking of the inversion symmetry in this geometry is important, as already pointed out above.

6

Accounting for Propagation Effects

For realistic calculations it is often not only necessary to know the nonlinear optical polarization but to actually consider the light field emitted by some gas or sample. This emission can also be influenced by group-velocity dispersion, reabsorption effects, reflections from interfaces and resulting standing-wave effects, phase matching, cascade processes, self-focusing or defocusing, diffraction, the Gouy phase, etc. Usually, exact analytical solutions are not available. In the following we first discuss two approaches in a one-dimensional plane-wave approximation: Exact numerical calculations using a finite-difference time-domain (FDTD) algorithm and the usual slowly varying envelope approximation (SVEA) [5,6], which can be somewhat generalized under certain conditions [154]. The SVEA is well known from traditional nonlinear optics, FDTD calculations are generally suitable for extreme nonlinear optics. Examples of one-dimensional FDTD calculations will be given in Sects. 7.1.2 and 7.2 for semiconductor experiments, phase-matching issues are, e.g., discussed in Sect. 7.5 for conical harmonic generation and in Sect. 8.1 for high-harmonic generation on the basis of the SVEA. Finally, we discuss the influence of the transverse (Gaussian) beam profile on the carrier-envelope offset phase of few-cycle pulses via the Gouy phase in Sect. 6.3. The implications for the amplitude spectrum and the pulse envelope are discussed in Sect. 6.4.

6.1 Numerical Solution of the Nonlinear Maxwell Equations

It is the aim of this brief section to show that – at least in one dimension – exact numerical solutions of the nonlinear Maxwell equations are actually rather simple. A more complete overview on numerical solutions of the Maxwell equations based on finite-difference time-domain (FDTD) algorithms can, e.g., be found in Ref. [155].

For a plane electromagnetic wave propagating along the z-direction with the electric field E polarized along x, the magnetic field B is directed along y. Under these conditions, Maxwell's equations (2.4) reduce to the two coupled first-order partial differential equations

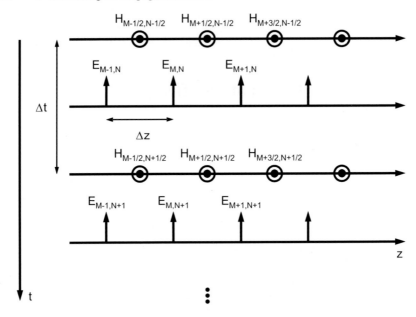

Fig. 6.1. Illustration of the one-dimensional finite-difference time-domain (FDTD) discretization and iteration scheme. The spatial step size is Δz, the temporal step size Δt, M and N are integers. The electric field E is parallel to the x-direction, the magnetic field H parallel to y (pointing out of the plane), and the wave is propagating along z.

$$\frac{\partial E(z,t)}{\partial z} = -\frac{\partial B(z,t)}{\partial t} \tag{6.1}$$

$$\frac{\partial H(z,t)}{\partial z} = -\frac{\partial D(z,t)}{\partial t} . \tag{6.2}$$

In a nonmagnetic medium we, furthermore, have $B(z,t) = \mu_0 H(z,t)$ and $D(z,t) = \varepsilon_0 E(z,t) + P(z,t)$, with the medium polarization $P(z,t)$.

In order to implement this *initial-value problem* on a computer, we discretize space and time according to

$$E_{M,N} = E(M\,\Delta z, N\,\Delta t) , \tag{6.3}$$

$$H_{M,N} = H(M\,\Delta z, N\,\Delta t) , \tag{6.4}$$

and P correspondingly. M and N are integers, the step sizes Δz and Δt have to be sufficiently small. Their actual choice will be discussed below. The central idea of Ref. [156] is to displace the positions of the electric and magnetic field by $\Delta z/2$ in space and $\Delta t/2$ in time, respectively (see Fig. 6.1). This leads to a simple iterative scheme: Replacing the partial derivatives in (6.1) by finite fractions immediately delivers

$$H_{M+\frac{1}{2},N+\frac{1}{2}} = H_{M+\frac{1}{2},N-\frac{1}{2}} - \frac{\Delta t}{\mu_0 \Delta z}\left(E_{M+1,N} - E_{M,N}\right) . \tag{6.5}$$

At some point in time, the initial value of the electric as well as magnetic field must be known at all positions z. If the magnetic field on the RHS of (6.5) is, e.g., known at time $t = (N - \frac{1}{2})\Delta t$ and the electric field is known at time $t = N\Delta t$, the magnetic field at time $t = (N + \frac{1}{2})\Delta t$ at all coordinates $z = (M + \frac{1}{2})\Delta z$ can directly be calculated from (6.5). Accordingly, (6.2) leads to

$$E_{M,N+1} = E_{M,N} \tag{6.6}$$
$$- \frac{\Delta t}{\epsilon_0 \Delta z}\left(H_{M+\frac{1}{2},N+\frac{1}{2}} - H_{M-\frac{1}{2},N+\frac{1}{2}}\right) - \frac{1}{\epsilon_0}\left(P_{M,N+1} - P_{M,N}\right).$$

On the RHS of (6.6), the magnetic field at time $t = (N + \frac{1}{2})\Delta t$ and the polarization and electric field at time $t = N\Delta t$ are known at this point of the calculation. $E_{M,N+1}$ and $P_{M,N+1}$ have to be calculated for all coordinates $z = M\Delta z$. In vacuum, $P = 0$ holds and $E_{M,N+1}$ can again be directly calculated. The same holds for a linear optical material with $P = \epsilon_0 \chi E$, in which case ϵ_0 has to be replaced by $\epsilon_0 \to \epsilon \epsilon_0$ and P disappears. In nonlinear optics, P is generally a nonlinear functional of E. If, for example, we have a $\chi^{(2)}$-medium with $P = \epsilon_0 \chi^{(2)} E^2$, we obtain $P_{M,N+1} = \epsilon_0 \chi^{(2)} E^2_{M,N+1}$, and (6.6) is a quadratic equation in $E_{M,N+1}$ allowing the electric field at time $t = (N + 1)\Delta t$ for all coordinates $z = M\Delta z$ to be determined. This then allows calculation of the magnetic field at time $t = (N + \frac{3}{2})\Delta t$ via (6.5), etc., which completes the iterative scheme. The knowledge of the electric field E and the magnetic field H delivers the Poynting vector $|S| = |E \times H| = EH$, hence also the light intensity I (see Sect. 2.2).

Typically, the step size Δz should be chosen smaller than one tenth of the smallest medium wavelength. Δt should be smaller than $\Delta z/v_{\mathrm{phase}}^{\max}$, where $v_{\mathrm{phase}}^{\max}$ is the largest (anticipated) phase velocity of the problem.

In practice, one has to be careful not to obtain artifacts from the spatial boundaries of the simulated region. The artificial reflections from these boundaries can, in principle, be delayed to very long times by making the simulated region sufficiently large. This "brute force" approach can, however, be rather CPU-time consuming. A faster and more elegant approach is to suppress these reflections by so-called absorbing boundary conditions [157] or by a projection-operator technique [158]. The latter has been used for some of the calculations presented in this book. It also allows for injection of the optical pulses from one side.

Complete FDTD solutions of the *three-dimensional* vectorial Maxwell equations for the focus of a laser in the regime of extreme nonlinear optics have not been published. They do, however, seem in reach with the computers at hand today.

6.2 Slowly Varying Envelope Approximation

In traditional nonlinear optics, the variations of the envelope of the optical polarization on the timescale of a period of light are small. This fact has already been the
basis for the rotating-wave approximation. Moreover, the variations of the optical polarization envelope are also usually small on the spatial scale of a wavelength of light
in traditional nonlinear optics – a statement that, however, depends on the density
of dipoles in the medium under consideration. This slow variation in both time and
space translates into a slow variation of the electric-field envelope versus time and
space according to the wave equation (2.10).

   Let us briefly recapitulate the resulting well-known form of the wave equation,
again in its one-dimensional version for propagation of a plane wave along the z-
direction. We introduce the ansatz

$$E(z,t) = \sum_{N=1}^{\infty} \frac{1}{2} \tilde{E}_N(z,t) \, e^{i(K_N z - N\omega_0 t - N\phi)} + \text{c.c.} \tag{6.7}$$

into (2.10) and replace $c_0 \rightarrow c(N\omega_0)$ there, which takes care of the linear part
of the optical polarization. Generally, this leads to an infinite set of coupled wave
equations. If, however, we assume that the N-th harmonic results primarily from the
fundamental laser field (and not from cascade processes, i.e., from a mixing of other
harmonics), and if we assume that the optical polarization can be expressed in terms
of the nonlinear optical susceptibilities according to (2.37) with the fundamental field
$E(z,t) = \tilde{E}_1 \cos(K_1 z - \omega_0 t - \phi)$, and if we assume that absorption of the fundamental
and the harmonics can be neglected, the situation simplifies drastically. Furthermore,
the spirit of the slowly varying envelope approximation specifically is to neglect as
many spatial and temporal derivatives of the envelopes $\tilde{E}_N$ as possible without getting
a trivial result. Accounting only for the first spatial derivative of the envelope $\tilde{E}_N$ on
the LHS of the wave equation and neglecting all temporal derivatives of the envelopes
on the RHS, this leads to

$$2 i K_N \, e^{i K_N z} \frac{\partial \tilde{E}_N}{\partial z}(z,t) = - \mu_0 \, (N\omega_0)^2 \, \epsilon_0 \tilde{\chi}^{(N)} \, \tilde{E}_1^N \, e^{i N K_1 z}, \tag{6.8}$$

with the dispersion relation

$$c(N\omega_0) = \frac{N\omega_0}{K_N}. \tag{6.9}$$

Here we have lumped all contributions of nonlinear susceptibilities $\chi^{(M)}$ (with $M = N, N+1, \ldots$) leading to a harmonic of order N into an effective susceptibility
$\tilde{\chi}^{(N)}$. Note that the CEO phase ϕ has dropped out. This is expected as the above
approximations imply that the spectral components of the various harmonics N must
not overlap in frequency space. Thus, a dependence on the CEO phase ϕ cannot occur.

   With the initial condition $\tilde{E}_N(z = 0, t) = 0$ for $N \geq 2$ and assuming that
the fundamental wave $\tilde{E}_1$ propagates undistorted and undepleted, the straightforward
solution of (6.8) leads to the intensity I_N of the N-th harmonic emitted from a medium
of thickness l for an incident laser intensity I according to (2.16)

$$I_N = \frac{1}{2} \sqrt{\frac{\varepsilon_0}{\mu_0}} \, |\tilde{E}_N(z = l, t)|^2$$

$$= \left(\mu_0 \frac{(N\omega_0)^2}{2K_N} \varepsilon_0 \tilde{\chi}^{(N)} \right)^2 \left(2\sqrt{\frac{\mu_0}{\varepsilon_0}} \right)^{N-1} I^N \, l^2 \, \text{sinc}^2 \left(\frac{\Delta K l}{2} \right), \quad (6.10)$$

with the sinc-function $\text{sinc}(X) = \sin(X)/X$ and the wavevector mismatch

$$\Delta K = NK_1 - 1K_N = N\omega_0 \left(\frac{1}{c(\omega_0)} - \frac{1}{c(N\omega_0)} \right). \qquad (6.11)$$

If the phase velocities $c(\omega_0)$ and $c(N\omega_0)$ are equal, we have $\Delta K = 0$, i.e. *phase matching*, and the harmonic intensity increases quadratically with the medium thickness l because $\text{sinc}(0) = 1$. If this condition is not fulfilled, the harmonic intensity first increases with l, but then it drops again for $\Delta K l \approx \pi$. For $\Delta K l = 2\pi$, it is strictly zero. This is a result of destructive interference between contributions of the N-th harmonic generated in front of the sample, which have propagated with phase velocity $c(N\omega_0)$ to the end of the sample on the one hand, and contributions generated at the end by a fundamental field that had to propagate with phase velocity $c(\omega_0)$ from the front towards the end on the other hand. This condition defines the *coherence length* of the N-th harmonic

$$l_{\text{coh}}(N) = \frac{\pi}{|\Delta K|}. \qquad (6.12)$$

In the semiconductor experiments on "THG in the disguise of SHG" discussed in this book is can be as low as just a few µm (Sect. 7.2), for high harmonics from gas jets it depends on the gas pressure and is typically several tens of µm (see Example 8.1). In gas-filled glass capillaries with tailored dispersion (Sect. 8.1.2), it may approach millimeters, for quasi phase-matching in modulated capillaries, it can effectively become centimeters (see Sect. 8.1.3).

6.3 Gouy Phase and Carrier-Envelope Phase

So far in this chapter, we have discussed propagation in a one-dimensional plane-wave approximation. In the focus of a lens, however, the transverse beam profile can also lead to interesting and relevant effects. Here, we only discuss the resulting modifications of few-cycle laser pulses, especially the influence on their CEO phase – which is relevant for extreme nonlinear optics – in terms of *linear optics*. Unlike in Sect. 2.3, where the variation of the CEO phase originated from material dispersion, the effects to be discussed now are purely topological, i.e., the variation of the CEO phase does also occur in vacuum. We first revisit the physics underlying the "well-known" Gouy phase and then discuss its implications on the CEO phase.

A reminder on the Gouy phase

As usual, when focusing a laser propagating along the z-direction with a lens or a spherical mirror, the transverse profile of the transverse component (e.g., the x-component) of the electric field at frequency ω for the fundamental Gaussian mode (see Fig. 6.2) in the Fresnel approximation follows [159]

$$E(r, t) = \frac{\tilde{E}_0}{2} \frac{w_0(\omega)}{w(z, \omega)} \exp\left(i \frac{|K|}{2} \frac{x^2 + y^2}{\mathcal{R}(z, \omega)}\right) e^{i(|K||z - \omega t - \varphi_G(z, \omega))} + \text{c.c.} \quad (6.13)$$

Here, the complex radius of curvature $\mathcal{R}(z, \omega)$ is related to the real radius of curvature $R(z, \omega)$ and the "width" $w(z, \omega)$ of the transverse Gaussian profile via

$$\frac{1}{\mathcal{R}(z, \omega)} = \frac{1}{R(z, \omega)} + i \frac{1}{w^2(z, \omega)} \frac{2}{|K|}, \quad (6.14)$$

with

$$R(z, \omega) = z + \frac{z_R^2(\omega)}{z} \quad (6.15)$$

and

$$w^2(z, \omega) = w_0^2(\omega) \left(1 + \left(\frac{z}{z_R(\omega)}\right)^2\right). \quad (6.16)$$

$w(0, \omega) = w_0(\omega)$ is the *beam waist* and

$$z_R(\omega) = \frac{w_0^2(\omega)}{2c_0} \omega \quad (6.17)$$

the *Rayleigh length* at frequency ω. The normalization in (6.13) has been chosen such that $\tilde{E}_0$ is the peak electric field in the focus at $r = 0$, which leads to

$$\int_{-\infty}^{+\infty} \int_{-\infty}^{+\infty} \langle E^2(r, t)\rangle \, dx \, dy = \frac{1}{4} \tilde{E}_0^2 \pi w_0^2(\omega) = \left(\frac{\tilde{E}_0}{2\sqrt{\ln(\sqrt{2})}}\right)^2 \pi r_{\text{HWHM}}^2 \quad (6.18)$$

for all z, where $\langle ... \rangle$ is again the cycle average. In the last step we have introduced the half-width at half-maximum (the "radius") of the Gaussian *intensity* profile r_{HWHM} for later use (see Sect. 7.1). Finally, and most importantly, the phase shift in (6.13)

$$\varphi_G(z, \omega) = \arctan\left(\frac{z}{z_R(\omega)}\right) \quad (6.19)$$

is the so-called *Gouy phase*, which has been known for more than a century [160, 161]. Still, it tends to remain a puzzle [162]. Intuitively, the narrow beam waist in the focus at $z = 0$ (see Fig. 6.2) corresponds to a broad distribution of transverse momenta of light via the "uncertainty" relation. As the modulus of the wavevector of light $|K| =$

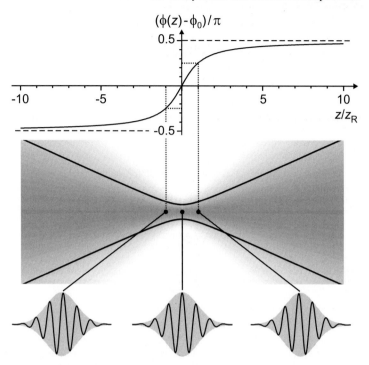

Fig. 6.2. Top: Variation of the CEO phase ϕ versus propagation coordinate z in units of the Rayleigh length $z_R(\omega_0)$ according to (6.22). $\phi_0 = \phi(0)$ is a phase offset. Middle: Scheme of the Gaussian beam profile according to (6.16) on the same scale. Bottom: Illustration of the resulting electric field $E(t) = \tilde{E}(t)\cos(\omega_0 t + \phi)$ for $\phi_0 = 0$ at the three on-axis positions $z = -z_R$, $z = 0$ and $z = +z_R$, respectively. Over $\pm$ one Rayleigh length, the CEO phase changes by $\pm\pi/4$. Such a $\pi/2$ change in phase can make all the difference (see, e.g., Figs. 5.2 and 5.6). The gray areas are the pulse envelope $\tilde{E}(t)$ [269].

$\sqrt{K_x^2 + K_y^2 + K_z^2} = \omega/c_0$ is constant for any given ω, this effectively reduces its z-component K_z. When the beam spreads in the transverse direction, while propagating along z, the transverse momentum distribution narrows down and $K_z(z)$ increases and finally approaches $K_z(z) = |K|$ for $|z| \to \infty$. The z-component, appropriately averaged [162] over the transverse beam profile, is called the propagation constant $K_z^{\mathrm{eff}}(z)$. Thus, on propagation from $-\infty$ to z, the wave acquires a phase relative to a plane wave given by

$$\int_{-\infty}^{z} \left(|K| - K_z^{\mathrm{eff}}(z') \right) \mathrm{d}z' = \varphi_G(z, \omega) . \tag{6.20}$$

Impact on the carrier-envelope phase

In other words: $\varphi_G(+\infty, \omega) - \varphi_G(-\infty, \omega) = \pi$ means that the phase of the wave propagates from $-\infty$ to $+\infty$ *faster* than a plane wave by half a cycle of light –

the phase velocity is superluminal, especially near to the focus[1]. The *average* group velocity, on the other hand, is luminal. It is again the difference between phase and group velocity that leads to a shift of the carrier-envelope offset phase ϕ (see Sect. 2.3).

In order to proceed, one first needs to specify the boundary conditions: To evaluate the CEO phase $\phi(z)$ via $\varphi_G(z, \omega)$ we need $z_R(\omega)$ for which we need $w_0(\omega)$, which depends on the width of the beam in front of the lens, $w(z_f, \omega) = w_f(\omega)$, at focal length $z_f \gg z_R(\omega_0)$. Here, we distinguish between the three cases summarized in Table 6.1.

Table 6.1. Overview of the three cases discussed in the following.

	Beam waist w_0	Beam at lens w_f	Rayleigh length z_R	In focal plane
(i)	$\propto 1/\omega$	$\propto$ const.	$\propto 1/\omega$	$v_{group} = v_{phase} \geq c_0$
(ii)	$\propto 1/\sqrt{\omega}$	$\propto 1/\sqrt{\omega}$	$\propto$ const.	$v_{group} = c_0 \leq v_{phase}$
(iii)	$\propto$ const.	$\propto 1/\omega$	$\propto \omega$	$v_{group} \leq c_0 \leq v_{phase}$

Case (i) corresponds to a frequency-independent beam width in front of the lens – an "ideal" laser beam [163]. The mode of *case (ii)* [164–166] has a frequency dependence, but when this mode spreads in size, the relative frequency distribution does not change – unlike in cases (i) and (iii) – because the Rayleigh length is frequency independent. While it may seem like a mathematical curiosity at first sight, it has been argued [166] that this case corresponds to the natural spatiotemporal mode of an open electromagnetic cavity: In order for a short pulse to periodically bounce back and forth between two spherical mirrors of a cavity, the radius of curvature R of *all* frequency components at the mirror has to match the radius of curvature of the mirror. Equation (6.15) leads to $z_R(\omega) = $ const. and with (6.17) we get the beam waist $w_0(\omega) \propto 1/\sqrt{\omega}$. The beam waist for *case (iii)* [167] does *not* depend on frequency – an "ideal" focus. If, for example, one sends a laser pulse through a spatial filter, a pinhole, and reimages this beam, one comes close to (iii).

To compute the variation of the on-axis ($x = y = 0$) electric field versus propagation coordinate z, we just have to sum (6.13) over the different frequency components. We obtain

$$E(0, 0, z, t) = \frac{1}{\sqrt{2\pi}} \int_0^\infty \frac{E_+(\omega)}{\sqrt{1 + \left(\dfrac{z}{z_R(\omega)}\right)^2}} \, e^{i(|K|z - \omega t - \varphi_G(z,\omega))} \, d\omega + \text{c.c.},$$

(6.21)

where $E_+(\omega)$ is the positive-frequency part of the Fourier transform of the electric field *in the focus* at $r = 0$. Here we have tacitly assumed that the decomposition

---

[1] Such superluminal phase velocity also occurs for a wave propagating through a subwavelength aperture ("photon tunneling") or through any empty waveguide close to its cutoff frequency.

into carrier wave and envelope remains meaningful (see Example 2.4). For $E(t) = \tilde{E}_0 \cos(\omega_0 t + \phi_0)$ we get $E(\omega) = E_+(\omega) + E_-(\omega) = \frac{\tilde{E}_0}{2}\sqrt{2\pi}\,(e^{-i\phi_0}\delta(\omega - \omega_0) + e^{+i\phi_0}\delta(\omega + \omega_0))$, which recovers (6.13) for $\omega = \omega_0$ and $\phi_0 = 0$. Generally, (6.21) has to be worked out numerically. For *case (ii)*, where $z_R(\omega) = z_R(\omega_0)$, thus $\varphi_G(z, \omega) = \varphi_G(z, \omega_0)$, it is immediately clear (see Sect. 2.3, Example 2.4, Problem 2.5, or the two-color case in Sect. 3.5) that the phase factor directly determines the CEO phase $\phi(z)$ of the pulse at position z, i.e.,

$$(\phi(z) - \phi_0) = \varphi_G(z, \omega_0) = \arctan\left(\frac{z}{z_R(\omega_0)}\right), \tag{6.22}$$

where $\phi_0 = \phi(0)$ is a phase offset that depends on the initial conditions. This dependence is illustrated in Fig. 6.2. Beautiful corresponding animations can be found in Ref. [168].

There is no terribly large difference for *cases (i) and (iii)*. Apart from reshaping of the pulse envelope via the amplitude spectrum (see next section), the Gouy phase for all ω, hence the CEO phase, is strictly the same for $z = -\infty$, 0, and $z = +\infty$ for (i)–(iii). Differences occur only in the precise shape of $(\phi(z) - \phi_0)$ in the focal region. To avoid numerical calculations at this point, we have a closer look at phase and group velocity, respectively. From (6.20) we get the general expression for the *phase velocity*

$$v_{\text{phase}}(z) = \frac{\omega}{K_z^{\text{eff}}} = \frac{\omega}{\dfrac{\omega}{c_0} - \dfrac{\partial}{\partial z}\varphi_G(z, \omega)}. \tag{6.23}$$

At the carrier frequency ω_0 and for all three cases, this leads to

$$v_{\text{phase}}(z) = \frac{c_0}{1 - \dfrac{c_0}{z_R(\omega_0)\,\omega_0}\mathcal{L}(z)} \geq c_0, \tag{6.24}$$

with the abbreviation

$$\mathcal{L}(z) = \frac{1}{1 + \left(\dfrac{z}{z_R(\omega_0)}\right)^2}, \tag{6.25}$$

a Lorentzian form with $\mathcal{L}(0) = 1$.

▶ **Example 6.1.** Suppose that we have a width of the beam in front of the lens of $w_f = 1$ mm, a focal length of $z_f = 10$ cm and a carrier wavelength of $\lambda = 0.8\,\mu\text{m} = 2\pi c_0/\omega_0$. This leads to a beam waist of $w_0 = 25.5\,\mu\text{m}$ and a Rayleigh length of $z_R(\omega_0) = 2.5$ mm. The phase velocity in the focus at $z = 0$ according to (6.24) is 5×10^{-5} above the vacuum speed of light, i.e., $v_{\text{phase}}/c_0 = 1.00005$.   ◀

The general expression for the *group velocity* is

$$v_{group}(z) = \frac{\partial \omega}{\partial K_z^{eff}} = \frac{1}{\dfrac{\partial}{\partial \omega}\left(\dfrac{\omega}{c_0} - \dfrac{\partial}{\partial z}\varphi_G(z, \omega)\right)} . \tag{6.26}$$

Upon working out the derivatives at the carrier frequency ω_0 we get

$$v_{group}(z) = \frac{c_0}{1 - \mathcal{F}\dfrac{c_0}{z_R(\omega_0)\,\omega_0}\,\mathcal{L}(z)\,(1 - 2\mathcal{L}(z))} , \tag{6.27}$$

with the factor $\mathcal{F} = -1, 0,$ and $+1$ for case (i), (ii), and (iii), respectively (also see RHS of Table 6.1). These dependences are illustrated in Fig. 6.3. For cases (i) and (iii), the group velocity crosses the vacuum velocity of light c_0 at the positions $z/z_R(\omega_0) = \pm 1$. Altogether, we expect that the slope of $\phi(z)$ in the focal plane at $z = 0$ is different for the three cases (i)–(iii). For cases (i) and (iii), the group velocity changes with z such that the group delay from $z = -\infty$ to $+\infty$ is strictly identical to that of a plane wave, which is luminal. This can be seen from the fact that all frequency components experience a π phase shift according to (6.19) when going from $z = -\infty$ to $+\infty$, which is simply a flip in sign of $E(t)$ – equivalent to a π shift of the CEO phase.

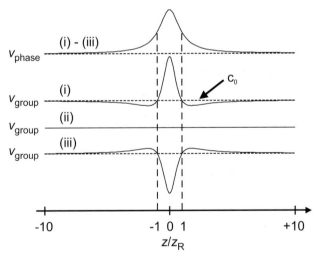

Fig. 6.3. Illustration of the phase and group velocities versus propagation coordinate z in units of the Rayleigh length $z_R(\omega_0)$ according to (6.24) and (6.27), respectively, for the three cases (i)–(iii) (see Table 6.1). The dashed horizontal lines indicate the vacuum velocity of light c_0, the maximum relative deviation from c_0 is given by the dimensionless parameter $c_0/(z_R(\omega_0)\,\omega_0) \ll 1$, see Example 6.1.

This π phase shift has directly been observed in the time domain for single-cycle THz pulses [169], also see Ref. [170]. In the optical regime, the variation $\phi(z)$ has recently been observed [171] in experiments employing above-threshold ionization of atoms (see Chap. 5). The variation of ϕ versus z has to be considered if a nonlinear medium, the thickness of which exceeds the Rayleigh length, is brought into the focus of a few-cycle pulse. It obviously leads to an averaging with respect to the CEO phase that smears out CEO-phase effects and, e.g., reduces the height of the peak at the carrier-envelope offset frequency f_ϕ in the RF spectrum. These effects of the Gouy phase are not very important for most of the solid-state experiments outlined in Chap. 7 as the samples there are much thinner than the Rayleigh length (with the notable exception of, e.g., the thick ZnO crystals in Sect. 7.2). They can be relevant for high-harmonic generation from gas jets (Sect. 8.1.1) or from relativistic nonlinear Thomson scattering (Sect. 8.2). In both cases, the extent of the gas jet is typically comparable to the Rayleigh length. A related effect also influences the phase-matching condition in gas-filled hollow waveguides (Sect. 8.1.2). In the language of the present section, the Gouy phase is simply linear in z in that case, equivalent to a z-independent propagation constant K_z^{eff} in Sect. 8.1.2 (see (6.20)). The variation $\phi(z)$ of the CEO phase also has to be considered if a laser pulse impinges onto a surface at an angle with respect to the surface normal as, e.g., for the photoemission from a metal surface (see Sect. 5.5). The difference in the fundamental Gouy phase and that of the N-th harmonic can also severely influence the phase-matching condition – even in traditional nonlinear optics [9].

The detailed theoretical description of actual experiments can be more involved than our treatment. Often, in order to get small foci, one overilluminates the aperture of the lens (or spherical mirror). Consequently, one has a *truncated* transverse Gaussian profile right after the lens [172] rather than a Gaussian according to (6.13). In this case the symmetry with respect to the focal plane at $z = 0$ is broken [172]. Also note that we have completely neglected effects of spherical or chromatic abberations of the lens.

6.4 Reshaping of the Amplitude Spectrum

We now have a closer look at the change of the amplitude spectrum via the $1/\sqrt{\cdots}$ factor in (6.21), which leads to a variation of the temporal envelope of the pulse versus z as well. Furthermore, one generally also gets a variation versus the radial coordinate $r = \sqrt{x^2 + y^2}$. We start with the latter aspect.

Case (i)

$w(z_f, \omega) = w_f$ at focal length $z_f \gg z_R(\omega_0)$ is frequency independent, hence we can write the beam waist as

$$w_0(\omega) = \frac{z_f\, 2c_0}{w_f\, \omega} = w_0(\omega_0)\, \frac{\omega_0}{\omega} =: w_0\, \frac{\omega_0}{\omega}. \tag{6.28}$$

Let us first consider the example of an incident Gaussian pulse. The positive-frequency part of the Gaussian laser spectrum in front of the lens on the optical axis be (see Example 2.4)

$$E_+(\omega) = \frac{\tilde{E}_0}{\sigma\sqrt{2}}\, e^{-\frac{(\omega-\omega_0)^2}{\sigma^2}}\, e^{-i\phi}, \tag{6.29}$$

corresponding to $E(t) = \tilde{E}(t)\cos(\omega_0 t + \phi)$ with envelope $\tilde{E}(t) = \tilde{E}_0 \exp(-(t/t_0)^2)$ in the time domain. $\sigma = 2/t_0$ is the spectral width. The spectrum in the focal plane $(z = 0)$ results as

$$E_+(r,\omega) = \frac{\tilde{E}_0}{\sigma\sqrt{2}}\frac{w_f}{w_0(\omega)}\, e^{-\frac{r^2}{w_0^2(\omega)}}\, e^{-\frac{(\omega-\omega_0)^2}{\sigma^2}}\, e^{-i(\phi+\frac{\pi}{2})} \tag{6.30}$$

$$= \frac{\tilde{E}_0}{\sigma\sqrt{2}}\frac{w_f\,\omega}{w_0\,\omega_0}\, e^{-\frac{(\omega-\tilde{\omega}_0(r))^2}{\tilde{\sigma}^2(r)}}\, \underbrace{\exp\left(-\frac{r^2}{w_0^2}\times\frac{1}{1+\frac{r^2\sigma^2}{w_0^2\omega_0^2}}\right)}_{\text{only a function of } r}\, e^{-i(\phi+\frac{\pi}{2})}.$$

Here we have accounted for the shift of the CEO phase via the Gouy phase. In the last step we have introduced the abbreviations

$$\frac{1}{\tilde{\sigma}^2(r)} = \frac{1}{\sigma^2}\left(1+\frac{r^2\sigma^2}{w_0^2\omega_0^2}\right) \tag{6.31}$$

and

$$\tilde{\omega}_0(r) = \frac{\omega_0}{1+\frac{r^2\sigma^2}{w_0^2\omega_0^2}} \le \omega_0. \tag{6.32}$$

Obviously, the *effective center frequency* $\tilde{\omega}_0(r)$ of the Gaussian shifts towards lower frequencies with increasing r. This effect is large for large values of σ^2/ω_0^2, i.e., it is important for short laser pulses. The fact that $E_+(r,\omega)\propto\omega$ somewhat counteracts this effect and tends to shift the spectrum towards higher frequencies on the optical axis. At the same time, the *effective spectral width* $\tilde{\sigma}(r)$ decreases when going away from the optical axis at $r=0$.

The quantity $|\sqrt{r}\,E_+(r,\omega)|$ is illustrated in Fig. 6.4 for a single-cycle optical pulse $\Leftrightarrow \sigma/\omega_0 = 2/(t_0\,\omega_0) = 2\sqrt{\ln\sqrt{2}}/\pi = 0.3748$ (compare Example 2.4). We choose to depict this quantity, because its square, integrated with respect to r in polar coordinates, is proportional to the total power spectrum. In other words: In this way one can assess the actual weight of the radial contributions. Figure 6.5 shows the same dependence for an incident octave-spanning box-shaped laser spectrum.

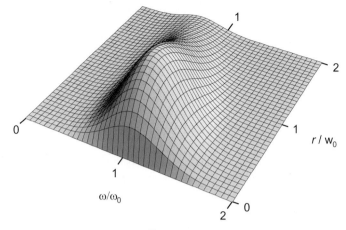

Fig. 6.4. Modulus of the electric field, $|\sqrt{r}\,E_+(r,\omega)|$, in the focal plane ($z = 0$) of a Gaussian beam profile of a Gaussian laser pulse versus spectrometer frequency ω in units of the laser carrier frequency ω_0 and radial coordinate r in units of the beam waist $w_0 = w_0(\omega_0)$. The parameter is $\sigma/\omega_0 = 0.3748$, equivalent to a single-cycle optical pulse. Note that the spectrum is slightly *blueshifted* with respect to $\omega/\omega_0 = 1$ at $r/w_0 = 0$, whereas it is significantly *redshifted* with respect to $\omega/\omega_0 = 1$ at $r/w_0 = 2$. The radial variation of the spectrum disappears for pulses containing many cycles of light and is already hardly visible for a two-cycle Gaussian pulse (not shown).

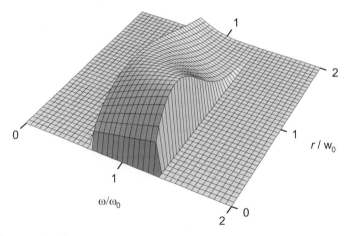

Fig. 6.5. Same as Fig. 6.4, but for an octave-spanning box-shaped incident spectrum. Here, the spectral reshaping is yet more obvious.

Intuitively, these fairly small effects again simply arise from the frequency dependence of diffraction. It is clear that the radial variation of the fundamental spectrum in the focal plane leads to a variation of the harmonics as well. This can also influence details of autocorrelation measurements of laser pulses or so-called frequency-resolved optical gating (FROG) – even if very thin SHG crystals are used.

Case (ii)

For a Gaussian pulse as in (i), one gets

$$\tilde{\sigma}(r) = \sigma \qquad (6.33)$$

and

$$\tilde{\omega}_0(r) = \omega_0 \left(1 - \frac{1}{2} \frac{r^2 \sigma^2}{w_0^2 \omega_0^2}\right), \qquad (6.34)$$

i.e., the qualitative behavior is the same as in (i). As the Rayleigh length is frequency independent, there is strictly no change of the shape of the on-axis electric field envelope or spectrum versus propagation coordinate z.

Case (iii)

A frequency-independent beam waist w_0 in the focus clearly means that the pulse envelope varies with the radial coordinate for the incident Gaussian laser beam in front of the lens (as for case (ii)). With this assumption, one eliminates (by construction) the radial variation of the pulse envelope in the focus, but the variation of the on-axis electric field along z remains (see $1/\sqrt{...}$ factor in (6.21)). This dependence is illustrated in Fig. 6.6 for an octave-spanning box-shaped spectrum in the focus at $r = 0$. Its low-frequency end is at $\omega = 2/3\,\omega_0$, its high-frequency end at $\omega = 4/3\,\omega_0$. Thus, over one Rayleigh length $z_R(\omega_0)$, the relative on-axis *intensity* changes by a factor of $(\sqrt{1 + (3/2)^2}/\sqrt{1 + (3/4)^2}\,)^2 = 52/25 \approx 2$ when going from the low- to the high-frequency end. Still, this hardly changes the FWHM of the temporal intensity

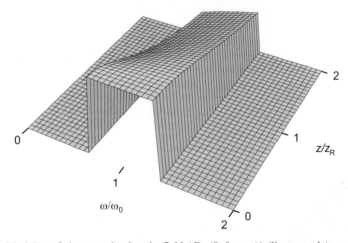

Fig. 6.6. Modulus of the on-axis electric field $|E_+(0, 0, z, \omega)|$ (linear scale) according to (6.21) versus propagation coordinate z in units of the Rayleigh length $z_R(\omega_0)$ and spectrometer frequency ω in units of the carrier frequency ω_0 for an octave-spanning box-shaped spectrum in the focus at $r = 0$.

profile (check this!) but it does change the shape of the pulse in its wings. For high-order processes, however, only the central part of the pulse really matters (see, e.g., Problem 2.3).

For a discussion of wave propagation including transverse effects in terms of an approximative scalar first-order propagation equation suitable for the regime of extreme nonlinear optics, we refer the reader to Ref. [135]. Such a treatment is relevant for high-harmonic generation from atoms.

7

Extreme Nonlinear Optics of Semiconductors and Isolators

In a solid, the atoms are arranged in the form of a periodic lattice with lattice constant a. The overlap of the electronic wave functions lifts the degeneracy of the discrete atomic energy levels and leads to energy bands. The wave functions become Bloch waves, which have a particular dispersion relation. Figure 7.1 schematically shows this dispersion relation for a direct-gap semiconductor with bandgap energy E_g. At zero temperature, the valence band is fully occupied and the conduction band is completely empty. Light can couple to the crystal electrons in two ways: via *interband* transitions and via *intraband* transitions.

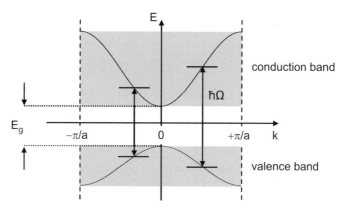

Fig. 7.1. Scheme of the valence and the conduction band of a direct gap semiconductor in the first Brillouin zone, i.e., for wave numbers, k, in the interval $[-\pi/a, +\pi/a]$, a is the lattice constant (actually, a tight-binding band structure is shown). At each k, the optical interband transition resembles that of a two-level system with transition energy $\hbar\Omega = E_c(k) - E_v(k)$. Close to the center of the Brillouin zone, the bands are nearly parabolic and the effective-mass approximation is justified. E_g is the bandgap energy.

Interband transitions

Light can promote an electron from an occupied state in the valence band into an empty state in the conduction band – provided the dipole moment is nonzero. Note that the photon wavelength (μm) is much larger than typical lattice constants ($a = 0.5$ nm). Consequently, the photon wavevector is a thousand times smaller than the electron wavevector at the edge of the first Brillouin zone, π/a, and such a transition looks very nearly vertical in the band structure. Thus, a state with wavevector k in the valence band is coupled to a state with the same wavevector k in the conduction band. Looking at these two states with transition frequency $\Omega(k)$ only, this problem is perfectly similar to what we have said about two-level systems in Sect. 3.2. On this level of description, semiconductor interband extreme nonlinear optics is just that of an ensemble of uncoupled two-level systems. Indeed, in Chap. 3 we have often plotted the emitted signals versus transition frequency Ω, allowing for applying these results to semiconductors as well. There are, however, deviations from this simplistic picture, which have to be discussed (Sect. 7.1.4). In any case, the important energy scale associated with the light field is the *Rabi energy* $\hbar\Omega_R$. If it becomes comparable with the transition energy $\hbar\Omega$, the regime of extreme nonlinear optics starts and effects like carrier-wave Rabi flopping (see Sect. 7.1) and "third-harmonic generation in the disguise of second-harmonic generation" (see Sect. 7.2) are expected.

In order to actually evaluate the peak Rabi energy $\hbar\Omega_R = d_{cv}\tilde{E}_0$, one needs to know the dipole matrix element for a transition from the valence (v) to the conduction (c) band at wavevector $\boldsymbol{k}$. Within $\boldsymbol{k} \cdot \boldsymbol{p}$ perturbation theory [22], the dipole matrix element d_{cv} is approximately wavevector independent and can be estimated on the basis of known material parameters by the following "rule of thumb"

$$|d_{cv}|^2 = \frac{\hbar^2 e^2}{2E_g}\left(\frac{1}{m_e} - \frac{1}{m_0}\right), \tag{7.1}$$

with the effective electron mass m_e and the free electron mass $m_0 = 9.1091 \times 10^{-31}$ kg.

It is interesting to relate the Rabi energy on the basis of (7.1) and the Keldysh parameter γ_K (see Sect. 5.2) from (5.9) at this point. Quoting a Keldysh parameter in this context means that we interpret a transition from the valence to the conduction band as laser-field-induced tunneling of a crystal electron through the bandgap, in which case the binding energy E_b in (5.9) is given by the bandgap energy E_g. This allows us to write

$$\frac{\hbar\Omega_R}{\hbar\omega_0}\gamma_K = \sqrt{1 - \frac{m_e}{m_0}}, \tag{7.2}$$

with the carrier photon energy $\hbar\omega_0$. Equation (7.2) can easily be verified by inserting Ω_R with d_{cv} from (7.1) and γ_K from (5.9) with $E_b \rightarrow E_g$ and assuming that d_{cv} is real. For GaAs parameters, $m_e/m_0 = 0.067$ and the square root on the RHS of (7.2) is $0.97 \approx 1$. This is also true for many other typical semiconductors, especially for those to be discussed below. We obtain the simple relation

$$\frac{\Omega_R}{\omega_0} \approx \frac{1}{\gamma_K},$$
(7.3)

which connects two seemingly rather different quantities, i.e. Ω_R and γ_K. Moreover, following Sect. 5.2 and (5.9), we can interpret the Rabi frequency $\Omega_R \approx \Omega_{tun}$ as the tunneling "frequency" Ω_{tun}. Indeed, we have already noticed in Sect. 5.2 that the two quantities, Ω_R and Ω_{tun}, are generally analogous.

Intraband transitions

The light field can also influence the states in the conduction (valence) band. Classically, this corresponds to the acceleration of an electron in the band. Near the center of the Brillouin zone, the bands can often be approximated by parabolas and the resulting physics is that of Volkov states (see Sect. 4.2). Extreme nonlinear optics starts if the Volkov sidebands acquire appreciable strength. This happens if the *ponderomotive energy* $\langle E_{kin} \rangle$ becomes comparable to the carrier photon energy $\hbar\omega_0$. The bands develop a series of sidebands, separated by $\hbar\omega_0$ (see Fig. 4.2). For a given laser intensity, the electron ponderomotive energy is usually much larger than the hole ponderomotive energy, because the electron effective mass is typically smaller than the hole effective mass by a factor of order ten. Inspection of Fig. 4.2 shows that a factor of ten in $\langle E_{kin} \rangle / \hbar\omega_0$ makes a very (!) large difference. We anticipate that the optical spectrum also develops sidebands separated by $\hbar\omega_0$, leading to oscillatory features in the optical spectrum. The low-energy sidebands lie in the gap of the semiconductor and lead to induced absorption below the gap. This altogether constitutes the *dynamic Franz–Keldysh effect*.

When do interband or intraband transitions, respectively, dominate?

Recall that the Rabi energy is proportional to the square root of the laser intensity I and that the ponderomotive energy is directly proportional to I. With increasing laser intensity, either of the two conditions $\hbar\Omega_R / \hbar\Omega \approx 1$ or $\langle E_{kin} \rangle / \hbar\omega_0 \approx 1$ is met first and, consequently, either interband or intraband transitions dominate the physics. Remember that the ratio $\langle E_{kin} \rangle / \hbar\omega_0$ is inversely proportional to ω_0^3. Thus, intraband effects are expected to dominate for infrared excitation, where $\hbar\omega_0 \approx 0.1\,eV$. The same scaling makes the ratio $\langle E_{kin} \rangle / \hbar\omega_0$ usually smaller than $\hbar\Omega_R / \hbar\Omega$ at accessible intensities and for optical frequencies ω_0, where $\hbar\omega_0 = 1.5$ to $3.0\,eV$. We will discuss both cases in this chapter.

Another facet related to intraband transitions and to the ponderomotive energy is the occurrence of *relativistic effects*. They become really large if the ponderomotive energy is comparable to the rest energy, i.e., if $|\mathcal{E}| = 2\sqrt{\langle E_{kin} \rangle / (m_e c_0^2)} \approx 1$ for crystal electrons with effective rest mass m_e. While this condition is typically only met at very much larger laser intensities than the above condition $\langle E_{kin} \rangle / \hbar\omega_0 \approx 1$, the associated effects, for example the *photon-drag effect*, are so unique and specific that they can be identified anyway – even if the electron velocities are not really relativistic yet. Again infrared excitation wavelengths are favored because the ratio $2\sqrt{\langle E_{\mathrm{kin}}\rangle/(m_{\mathrm{e}}c_0^2)}$ scales inversely with the carrier frequency ω_0 itself.

Apart from these examples, which can be understood in terms of noninteracting electrons, *excitons* can exhibit a physics closely similar to what we have said about extreme nonlinear optics of atoms in Chap. 5. After an interband transition, the optically generated, negatively charged electron in the conduction band and the remaining positively charged hole in the valence band attract each other via the Coulomb interaction, which can form a bound state similar to that of an electron in a hydrogen atom. This bound state is called an exciton. While the equations are identical to those introduced in Chap. 5, the numbers are very different. Excitons in semiconductors have typical binding energies E_{b} on the order of ten meV as opposed to ten eV for atoms. Thus, unity exciton Keldysh parameter $\gamma_{\mathrm{K}}^{\mathrm{x}}$ is reached at fairly low intensities for infrared excitation. This mechanism competes with the above Volkov sidebands in the infrared regime.

7.1 Carrier-Wave Rabi Flopping

The phenomenon of carrier-wave Rabi flopping refers to Rabi flopping under conditions where the Rabi frequency Ω_{R} becomes comparable with the carrier frequency ω_0 of the exciting laser pulses. Rabi flopping occurs for resonant excitation of a two-level system with transition frequency Ω, i.e., for $\Omega/\omega_0 = 1$. We will see below that the transitions near the semiconductor bandgap dominate the nonlinear optical response. Thus, the model semiconductor GaAs with a bandgap energy of $E_{\mathrm{g}} = 1.42\,e\mathrm{V}$ is well suited for such experiments using pulses from a Ti:sapphire laser with $\hbar\omega_0 \approx E_{\mathrm{g}}$. The bandgap energy of $1.42\,e\mathrm{V}$ translates into a light period of 2.9 fs. The anticipated signatures of carrier-wave Rabi flopping have already been discussed in Sect. 3.3: If the Rabi energy becomes comparable to the carrier photon energy $\hbar\omega_0 = \hbar\Omega$, the third-harmonic peak splits into the carrier-wave Mollow triplet, consisting of a set of peaks at the three energies 3Ω, $3\Omega+\Omega_{\mathrm{R}}$ and $3\Omega-\Omega_{\mathrm{R}}$. For not too large Rabi energies, only the outer two peaks are visible. With increasing Rabi energy, they separate more and more, until the $3\Omega-\Omega_{\mathrm{R}}$ peak of the third harmonic meets the $\Omega+\Omega_{\mathrm{R}}$ peak of the fundamental Mollow triplet at the spectral position of the second harmonic at a Rabi energy given by $\Omega_{\mathrm{R}}/\Omega = \Omega_{\mathrm{R}}/\omega_0 = 1$. The interference of the two contributions with phase ϕ and 3ϕ, respectively, can lead to a dependence on the carrier-envelope offset phase ϕ.

Laser systems

However, the condition Rabi frequency Ω_{R} equal to the carrier frequency of light ω_0 corresponds to a large intensity by solid-state standards. While it is possible to reach this condition with pulses of several tens of femtoseconds in duration in terms of available lasers, it is not very likely that the semiconductor samples will survive the large deposited energy (= intensity × duration). Thus, it is favorable to study excitation with very short pulses, ideally with only one or two cycles of light in

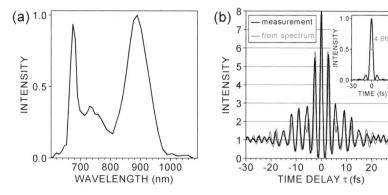

Fig. 7.2. Experiment: **(a)** measured laser spectrum, **(b)** measured interferometric autocorrelation. The gray curve in (b) is the autocorrelation computed from the laser spectrum under the assumption of a constant spectral phase (no chirp). The inset in (b) depicts a 4.8-fs full width at half-maximum real-time intensity profile computed under the same assumption. Figure reprinted from Ref. [54] by permission of O. D. Mücke.

duration with minimum deposited energy. For a discussion on laser-induced damage of dielectrics see Refs. [173–176].

These criteria can be met by excitation with 5-fs linearly polarized optical pulses directly from a mode-locked laser oscillator operating at $f_r = 81\,\text{MHz}\ (= 1/12\,\text{ns})$ repetition frequency [177]. The typical average output power of such a laser lies in the range from 120 to 230 mW. Figure 7.2(a) shows a typical laser spectrum, which has been obtained via Fourier transform of an interferogram taken with a pyroelectric detector, which is spectrally extremely flat. The Michelson interferometer used at this point is carefully balanced and employs beam splitters fabricated by evaporating a thin film of silver on a 100-µm thin glass substrate. The Michelson interferometer is actively stabilized by means of the Pancharatnam screw [178], which allows for continuous scanning of the time delay while maintaining active stabilization. The remaining fluctuations in the time delay between the two arms of the interferometer are around ±0.05 fs. The spectral wings that can be seen in Fig. 7.2(a) result from the spectral characteristics of the laser output coupler. The measured interferometric autocorrelation depicted in Fig. 7.2(b) is very nearly identical to the one computed from the spectrum (Fig. 7.2(a)) under the assumption of a constant spectral phase. This shows that the pulses are nearly transform limited. The intensity profile computed under the same assumption is shown as an inset in Fig. 7.2(b) and reveals a duration of about 5 fs. As a result of the strongly structured spectrum (a square-function to zeroth order), the intensity envelope versus time shows satellites. Using a high numerical aperture reflective-microscope objective [179], these pulses can be focused tightly to a profile that is very roughly Gaussian with $r_{\text{HWHM}} = 1\,\mu\text{m}$ radius (defined in (6.18)), as measured by a knife-edge technique at the sample position (see Fig. 7.5(a)). This sample position is equivalent to that of the second-harmonic generation (SHG) crystal used for the autocorrelation in terms of group-delay dispersion. From the known peak laser intensity I, the peak of the electric-field envelope $\tilde{E}_0$ can be calculated with (2.16). To estimate the Rabi frequency and/or the envelope pulse area $\tilde{\Theta}$, one, furthermore, needs the dipole matrix element d_{cv} of the optical transition from the valence (v) to the conduction (c) band. From the literature for GaAs one finds $d_{cv} = 0.3\,e\,nm$ [180] and $d_{cv} = 0.6\,e\,nm$. $\mathbf{k} \cdot \mathbf{p}$ perturbation theory according to (7.1) for GaAs parameters ($E_g = 1.42\,eV$ and $m_e = 0.067 \times m_0$) delivers $d_{cv} = 0.65\,e\,nm$. In what follows, $d_{cv} = 0.5\,e\,nm$ is used for GaAs.

Samples

Experiments have been performed on different types of GaAs samples. One is a 0.6-μm thin film of GaAs clad between $Al_{0.3}Ga_{0.7}As$ barriers, grown by metalorganic vapor phase epitaxy on a GaAs substrate. The sample is glued onto a 1-mm thick sapphire disk and the GaAs substrate is removed. Finally, a $\lambda/4$-antireflection coating is evaporated. This sample design [181] guarantees extremely high quality of the GaAs film. We will see below, however, that the experiments on this sample suffer from propagation effects to a certain extent. The other set of samples consists of a GaAs layer of thickness l (25, 50 and 100 nm) directly grown on a sapphire substrate in a molecular-beam epitaxy machine [182]. Although these samples are not comparable in linewidth with the above $GaAs/Al_{0.3}Ga_{0.7}As$ double heterostructure, the $l = 100$-nm thin GaAs film on sapphire substrate does exhibit a band edge in linear optical transmission experiments at room temperature (not shown). As the relevant energy scales in these experiments are larger than 0.1eV anyway, linewidth is not much of an issue. A relevant drawback of this sample geometry is that surface SHG effects play a certain role.

7.1.1 Experiment

Let us first discuss results for single pulses exciting the GaAs double heterostructure. The light emitted by the sample is prefiltered and spectrally dispersed in a conventional grating spectrometer. Figure 7.3 shows spectra at the third harmonic of the GaAs bandgap for different pulse intensities I in multiples of I_0. Here, I_0 is defined as $I_0 = 0.6 \times 10^{12}\,W/cm^2$. One complete Rabi flop is expected [181] for $I = 0.601 \times I_0$. One Rabi flop within two optical cycles clearly corresponds to $\Omega_R/\omega_0 = 1/2$. At low intensity, i.e., for $I = 0.017 \times I_0$, we observe a single maximum around 300 nm wavelength that is interpreted as the usual third-harmonic generation, resonantly enhanced by the GaAs band edge. With increasing intensity, this single peak splits and a second maximum is observed emerging on the long-wavelength side. This splitting qualitatively resembles the outer two peaks of the third-harmonic carrier-wave Mollow triplet discussed in Sect. 3.3. We will see below that the central peak of this triplet is hardly visible under these conditions in theory as well.

   To get some insight into the underlying dynamics, additional experiments with pairs of pulses are interesting. Such pairs can be obtained from the same balanced Michelson interferometer that is used to record the interferometric autocorrelation. The time delay between these two pulses is called τ. Notably, the envelope pulse area

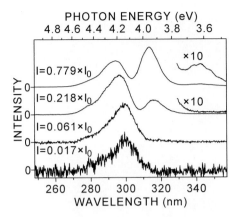

Fig. 7.3. Experiment: Spectra of light emitted into the forward direction around the third harmonic of the GaAs bandgap frequency. The spectra are shown on a linear scale, vertically displaced and individually normalized (from top to bottom: maxima correspond to 5664, 439, 34 and 4 counts/s). Excitation with 5 fs pulses. The intensity I of the pulses is indicated. Reprinted with permission from O. D. Mücke et al., Phys. Rev. Lett. **87**, 057401 (2001) [181]. Copyright (2001) by the American Physical Society.

$\tilde{\Theta}$ is the same for $\tau = 0$ and, e.g., for τ equal to two optical cycles – because the two optical fields simply add. Yet, the corresponding Rabi frequency is larger for $\tau = 0$ by about a factor of two. For low intensities (Fig. 7.4(a)), i.e., for small Rabi frequency as compared to the light frequency, the third-harmonic spectrum is simply modulated as a function of τ due to interference of the laser pulses within the sample leading to a period of about 2.9 fs. In contrast to this, for higher intensities (Fig. 7.4(b)–(d)), where the Rabi frequency becomes comparable to the light frequency, the shape of the spectra changes dramatically with time delay τ. For example, for $\tau = 0$ in Fig. 7.4(b), the two pulses simply interfere constructively and the same spectral double-maximum structure as in the single-pulse experiments (Fig. 7.3) is found. For larger τ, i.e., after one or two optical cycles, this double maximum disappears and is replaced by one prominent and much larger maximum. For the highest intensity, i.e., for Fig. 7.4(d) – which corresponds to an estimated envelope pulse area $\tilde{\Theta}$ of more than 4π – the behavior is quite involved with additional fine structure for $|\tau| < 1$ fs. Note that the spectra for $\tau = 0$ nicely reproduce the behavior seen in Fig. 7.3.

Further experiments have been performed, deliberately introducing positive or negative group-velocity dispersion by moving one of the extracavity CaF_2 prisms in or out of the beam with respect to the optimum position [181]. Obviously, this leaves the amplitude spectrum of the laser pulses unaffected. One quickly gets out of the regime of carrier-wave Rabi flopping, i.e. both, the splitting at $\tau = 0$ as well as the dependence of the shape on the time delay τ, quickly disappear with increasing pulse chirp. This demonstrates that it is not just the large bandwidth of the pulses but the fact that they are short and intense, which is important for the observation of carrier-wave Rabi flopping.

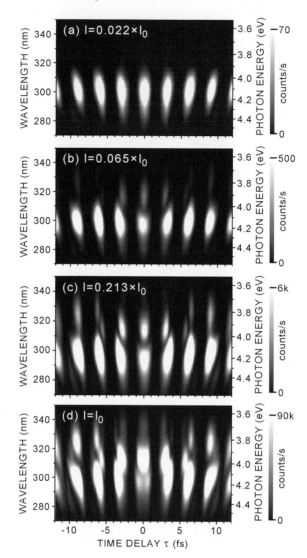

Fig. 7.4. Experiment: Same as Fig. 7.3, however, using pairs of phase-locked 5-fs pulses. The signal around the third harmonic of the bandgap is depicted versus time delay τ in a grayscale plot (note the saturated grayscale on the right hand side). **(a)** – **(d)** correspond to different intensities I as indicated. I refers to one arm of the interferometer. Reprinted with permission from O. D. Mücke et al., Phys. Rev. Lett. **87**, 057401 (2001) [181]. Copyright (2001) by the American Physical Society.

It is clear that one also expects large induced transparency at the GaAs bandgap as a result of the Rabi flopping. Figure 7.5(a) schematically shows the experimental geometry. To vary the excitation intensity without having to introduce filters, the sample is moved along the z-direction through the fixed focus ($z = 0$) of the microscope

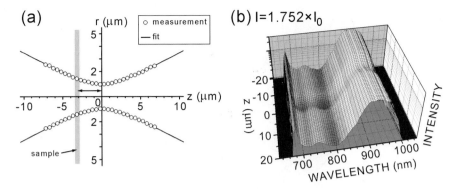

Fig. 7.5. (a) Scheme of the z-scan experiment (the measured radii are fitted with the formula $r(z) = r_{HWHM} \sqrt{1 + (z/z_R)^2}$ with $r_{HWHM} = 0.97\,\mu m$ and Rayleigh length $z_R = 2.9\,\mu m$, see Sect. 6.3), **(b)** transmitted light intensity on a logarithmic scale versus sample position z for an intensity of $I = 1.752 \times I_0$ (referring to $z = 0$). Figure reprinted from Ref. [54] by permission of O. D. Mücke.

objective. To enhance the visibility of the changes in transmission shown in Fig. 7.5, the differential transmission, $\Delta T / T$, is defined as

$$\frac{\Delta T}{T} = \frac{I_t(z) - I_t(z = -\infty)}{I_t(z = -\infty)}, \tag{7.4}$$

where $I_t(z)$ is the transmitted light intensity at sample position z. The condition $z = -\infty$ actually corresponds to $z = -20\,\mu m$ in the experiment, where the profile is so large that one can safely assume that linear optics applies.

Figure 7.6 shows corresponding results for three different incident light intensities I in units of I_0 as defined above. First, all results are closely symmetric around $z = 0$, which indicates that changes in absorption dominate. Changes in the refractive index might lead to focusing or defocusing of the beam that would result in an asymmetric dependence on z (similar to the known so-called z-scan technique, e.g., described in Ref. [183]). Secondly, one can see a large increase in transmission for wavelengths shorter than the GaAs band edge (approximately 870 nm) around $z = 0$ (Fig. 7.6(a)). $z = 0$ corresponds to the highest intensity in each plot. The maximum around 670 nm wavelength results from bleaching of the bandgap of the $Al_{0.3}Ga_{0.7}As$ barriers of the GaAs double heterostructure that accidentally coincides with the pronounced maximum in the laser spectrum (Fig. 7.2(a)) also around 680 nm. For larger intensity, Fig. 7.6(b), the transmission maximum around $z = 0$ flattens and we observe pronounced induced absorption for wavelengths longer than the GaAs band edge. For the highest intensity (Fig. 7.6(c)), this increased absorption becomes the dominating feature throughout most of the spectral range. Note that little if any induced transparency is observed for wavelengths between 780 nm (170 meV above the unrenormalized bandgap $E_g = 1.42\,eV$) and 700 nm (350 meV above the unrenormalized bandgap) while the laser spectrum (Fig. 7.2) still has significant amplitude

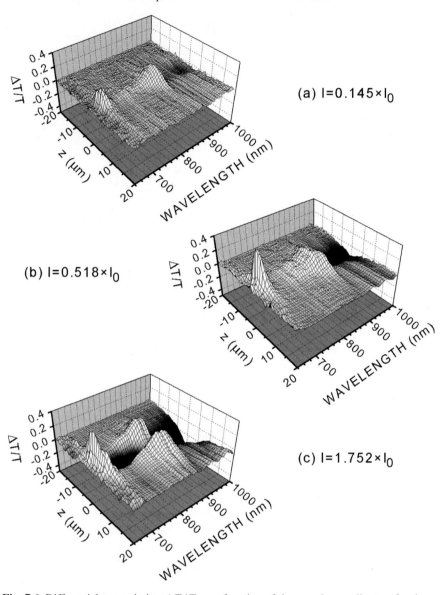

Fig. 7.6. Differential transmission $\Delta T/T$ as a function of the sample coordinate z for three different incident intensities I (referring to $z = 0$). **(a)** $I = 0.145 \times I_0$, **(b)** $I = 0.518 \times I_0$, **(c)** $I = 1.752 \times I_0$. Figure reprinted from Ref. [54] by permission of O. D. Mücke.

there. This indicates that these states high up in the band-to-band continuum of GaAs must experience a much (!) stronger damping (phase relaxation) and/or energy relaxation than the states near the bandgap under these conditions. This finding is an important input for the theoretical modeling.

7.1.2 Theory

Before proceeding to further experiments, we want to see how far theory for a semi-conductor is able to reproduce these experimental findings. Neglecting the Coulomb interaction of carriers, any type of intraband optical processes at this point, phonons and their coupling to the carriers, suppressing spin indices and using the dipole approximation for the optical transitions from the valence (v) to the conduction (c) band at electron wavevector k we have [22] , we have the Hamiltonian

$$\mathcal{H} = \sum_k E_c(k) \, c_{ck}^\dagger c_{ck} + \sum_k E_v(k) \, c_{vk}^\dagger c_{vk} \qquad (7.5)$$
$$- \sum_k d_{cv}(k) E(r, t) \left(c_{ck}^\dagger c_{vk} + c_{vk}^\dagger c_{ck} \right) .$$

Here $E_{c,v}(k)$ are the single-particle energies of electrons in the conduction and valence band, respectively (the band structure), which are schematically shown in Fig. 7.1. $d_{cv}(k)$ is the (real) dipole matrix element for an optical transition at electron wavevector k. The creation $c^\dagger$ and annihilation c operators create and annihilate crystal electrons in the indicated band (c,v) at the indicated momentum (k). The optical polarization is given by

$$P(r, t) = \frac{1}{V} \sum_k d_{cv}(k) \; (p_{vc}(k) + \text{c.c.}) + P_b(r, t), \qquad (7.6)$$

where the optical transition amplitudes

$$p_{vc}(k) = \langle c_{vk}^\dagger c_{ck} \rangle \qquad (7.7)$$

depend on time t as well as parametrically on the spatial coordinate r. As usual, the sum in (7.6) can be expressed via the combined density of states $D_{cv}(E)$ as $\sum_k \dots \to \int D_{cv}(E)\dots dE \to \sum_n D_{cv}(E_n)\dots \Delta E$, which neglects all anisotropies. Sometimes, the background polarization $P_b(r, t) = \varepsilon_0 \chi_b(z) E(r, t) = \varepsilon_0(\varepsilon_b(z) - 1)E(r, t)$ is employed, which approximately accounts for all "very" high-energy optical transitions not explicitly accounted for in (7.5). It can be expressed in terms of the background dielectric constant $\varepsilon_b(r)$. We will not use it in this section.

The dynamics of $p_{vc}(k)$, as well as those of the occupation numbers in the conduction band

$$f_c(k) = \langle c_{ck}^\dagger c_{ck} \rangle \qquad (7.8)$$

and in the valence band

$$f_v(k) = \langle c_{vk}^\dagger c_{vk} \rangle \qquad (7.9)$$

are easily calculated from the Heisenberg equation of motion for any operator $\mathcal{O}$ according to (3.15). Employing the usual anticommutation rules, i.e.

$$[c_{ck}, c_{ck'}^\dagger]_+ = \delta_{kk'} , \qquad [c_{vk}, c_{vk'}^\dagger]_+ = \delta_{kk'} , \qquad (7.10)$$

and that all other anticommutators are zero, this leads to the well-known semiconductor Bloch equations *for the transition amplitude*

$$\left(\frac{\partial}{\partial t} + i\Omega(\boldsymbol{k})\right) p_{vc}(\boldsymbol{k}) + \left(\frac{\partial}{\partial t} p_{vc}(\boldsymbol{k})\right)_{rel} \tag{7.11}$$

$$= i\hbar^{-1} d_{cv}(\boldsymbol{k}) E(\boldsymbol{r}, t) (f_v(\boldsymbol{k}) - f_c(\boldsymbol{k})) ,$$

with the *optical transition energy*

$$\hbar\Omega(\boldsymbol{k}) = E_c(\boldsymbol{k}) - E_v(\boldsymbol{k}) , \tag{7.12}$$

and *for the occupation in the conduction band*

$$\frac{\partial}{\partial t} f_c(\boldsymbol{k}) + \left(\frac{\partial}{\partial t} f_c(\boldsymbol{k})\right)_{rel} = 2\hbar^{-1} d_{cv}(\boldsymbol{k}) E(\boldsymbol{r}, t) \operatorname{Im}(p_{vc}(\boldsymbol{k})) . \tag{7.13}$$

Here we have again assumed a real dipole matrix element. $(1 - f_v(\boldsymbol{k}))$ can be interpreted as the occupation of holes and obeys an equation similar to $f_c(\boldsymbol{k})$. The terms with subscript "rel" have been added phenomenologically and describe dephasing and relaxation, respectively. For a state-of-the-art description of scattering processes see Refs. [22, 184–190]. For very short timescales, they are not too important. Note that the transition amplitude $p_{vc}(\boldsymbol{k})$ and the occupation factors $f_c(\boldsymbol{k})$ and $f_v(\boldsymbol{k})$ are easily connected to the components of the Bloch vector $(u, v, w)^T$, which we have extensively used in Chap. 3, via

$$\begin{pmatrix} u \\ v \\ w \end{pmatrix} := \begin{pmatrix} 2\operatorname{Re}(p_{vc}(\boldsymbol{k})) \\ 2\operatorname{Im}(p_{vc}(\boldsymbol{k})) \\ f_c(\boldsymbol{k}) - f_v(\boldsymbol{k}) \end{pmatrix} , \tag{7.14}$$

with the equation of motion (3.17) $(d \to d_{cv})$ and the *Rabi frequency* $\Omega_R(t)$ with

$$\hbar\Omega_R(t) = d_{cv} E(\boldsymbol{r}, t) . \tag{7.15}$$

For semiconductors in general, one gets additional contributions to the Rabi frequency as a result of the Coulomb interaction of electrons and holes [22, 191–198]. This modified or renormalized Rabi frequency can be interpreted as an internal field, which adds to the external laser field. Furthermore, the Coulomb interaction leads to scattering and dephasing as well as to energetic shifts of the states in the bands. Whether or not these aspects are relevant depends on the problem under consideration. An overview in terms of traditional nonlinear optics can be found in the textbooks [22, 199].

Let us first connect to the calculations shown in Sect. 3.3 by repeating them for parameters more suitable to the above GaAs experiments (Fig. 7.7). Here, ω again denotes the spectrometer frequency, ω_0 the laser carrier frequency and $\Omega = \Omega(\boldsymbol{k}) = \hbar^{-1}(E_c(\boldsymbol{k}) - E_v(\boldsymbol{k}))$ the transition frequency of one transition within the band. All

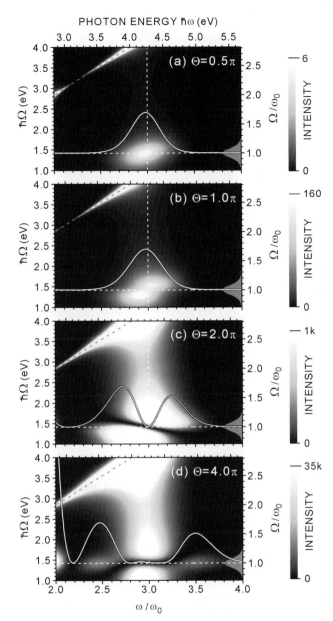

Fig. 7.7. Theory: Grayscale plot of the intensity (square modulus of p_{vc}) as a function of spectrometer frequency ω and transition frequency Ω (see Fig. 7.1). The carrier frequency ω_0 of the optical pulses (see gray areas on the RHS) is centered at the bandgap frequency, i.e., $\hbar\omega_0 = E_g$. The spectrum for a transition right at the bandgap, i.e., $\hbar\Omega = E_g$, is highlighted by the white curve. The diagonal dashed line corresponds to $\Omega = \omega$. Excitation with sech^2-shaped 5-fs pulses. The envelope pulse area $\tilde{\Theta}$ increases from **(a)** to **(d)** [269].

states are assumed to have the same dipole matrix element and the same phenomeno-logical relaxation in (7.11) according to

$$\left(\frac{\partial}{\partial t} p_{vc}(\boldsymbol{k})\right)_{rel} = \frac{p_{vc}(\boldsymbol{k})}{T_2}, \tag{7.16}$$

with the dephasing time $T_2 = 50\,\text{fs}$. Relaxation for the occupation numbers is ne-glected. Without bandgap renormalization, it is clear that there are no states below the bandgap energy (dashed horizontal line); nevertheless, we depict these data. Again, the laser carrier frequency is centered at the bandgap energy, i.e., we have $\hbar\omega_0 = E_g$. The laser spectrum is shown on the right-hand side lower corner as the gray-shaded area. The spectrum for $\hbar\omega_0 = \hbar\Omega = E_g$ is also depicted by the white line. It corre-sponds to the result of a single resonantly excited two-level system. For small envelope pulse area, $\tilde{\Theta} = 0.5\pi$, we find a single rather narrow maximum around $\omega/\omega_0 = 3$ and $\Omega/\omega_0 = 1$. Its width correlates with the width of the laser spectrum. This single maximum is just the usual, yet resonantly enhanced, third-harmonic generation. It experiences a constriction for $\tilde{\Theta} = 1.0\pi$, which evolves into a shape that resem-bles an anticrossing for $\tilde{\Theta} = 2.0\pi$. Here, two separate peaks are only observed in a rather narrow region around $\hbar\Omega = \hbar\omega_0 = E_g$, while for larger $\hbar\Omega$ only a single maximum occurs. Also, we find that the contribution of larger frequency transitions is by no means small. For example, for $\hbar\Omega = 2\,\text{eV}$ transition energy, the signal is actually larger than for the bandgap, i.e., for $\hbar\Omega = 1.42\,\text{eV}$. This trend continues for yet larger pulse areas (see $\tilde{\Theta} = 4.0\pi$ in Fig. 7.7(d)). While there is considerable resonant enhancement (as can be seen from Fig. 7.7(a)), this enhancement becomes less important at large pulse areas because the resonant transitions are completely saturated.

The actual spectra are the integral over the individual contributions, multiplied by the combined density of states, over the relevant range of transition energies. The bands themselves clearly have contributions even at $\hbar\Omega = 5\,\text{eV}$. If one would sum up all these contributions at, e.g., $\tilde{\Theta} = 4.0\pi$ (Fig. 7.7(d)), one no longer gets two maxima but rather a single maximum around $\omega/\omega_0 = 3$, which would no longer be in agreement with the experiments. The above transmission experiments (Fig. 7.6) have, however, already suggested that the dephasing/relaxation of the high-energy states is much faster. Thus, their nonlinear response is suppressed. On the low-energy end one has to account for the fact that bandgap renormalization becomes significant at these very large carrier densities. If one, e.g., integrates the spectra from 1.2 to 1.6 eV transition energy $\hbar\Omega$ with a constant density of states (not shown), the experimental behavior is reproduced quite well. In particular, one gets a gradual growth of a second spectral maximum rather than the sudden splitting observed for a single two-level system.

Another crucial aspect is to check for the importance of propagation effects (see Chap. 6). In order to do this in a realistic manner, an ensemble of two-level systems is employed, which fits the known and measured shape of the linear dielectric function of GaAs according to Ref. [201] (reproduced in Fig. 7.8(d)). This means that the high-frequency transitions are *not* treated via a background dielectric constant, which would correspond to an instantaneous response but rather is their finite response time

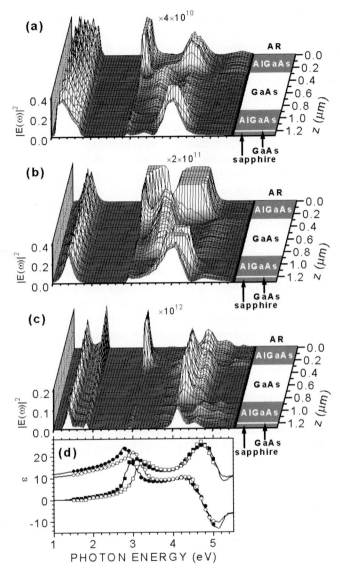

Fig. 7.8. $|E(\omega)|^2$ (normalized to the maximum of the incident electric field spectrum) as a function of $\hbar\omega$ and z, $\phi = 0$. Note the strong variation as a function of z. **(a)** sinc^2-shaped 5.6-fs pulses, $\tilde{E}_0 = 3.5 \times 10^9$ V/m, the GaAs cap layer thickness is $d_{cap} = 30$ nm. **(b)** As (a), but for sech^2-shaped 5-fs incident optical pulses with $\tilde{E}_0 = 3.5 \times 10^9$ V/m, and $d_{cap} = 10$ nm. **(c)** As (a), but for incident pulses that are fitted [200] to the experiment (see Fig. 7.2(a)), $\tilde{E}_0 = 2.5 \times 10^9$ V/m, $d_{cap} = 10$ nm. **(d)** The real (circles) and imaginary (squares) part of the linear dielectric function of GaAs (full) and $Al_{0.3}Ga_{0.7}As$ (open), respectively, are shown for comparison. The symbols are the experimental data from Ref. [201], the full curves correspond to our modeling. Reprinted with permission from O. D. Mücke et al., Phys. Rev. Lett. **89**, 127401 (2002) [202]. Copyright (2002) by the American Physical Society.

properly accounted for. The linear dielectric function of GaAs exhibits two strong resonances, the so-called E_1 and E_2 resonance, which are due to the particular shape of the band structure. In substantial regions of the Brillouin zone the bands move in parallel, which leads to a large combined density of states. Following our above discussion on high-energy transitions, only the optical nonlinearities of the band-edge transitions are accounted for, the other transitions are assumed to be linear. Mathematically, this is accomplished by setting their occupation to zero. The corresponding coupled Maxwell–Bloch equations in one dimension with phenomenological dephasing rates are solved numerically without further approximations [202]. This is a rather demanding numerical task. Results for the GaAs double heterostructure discussed above are shown in Fig. 7.8 for (a) 5.6-fs sinc²-shaped pulses, (b) 5-fs sech²-shaped pulses, and (c) for pulses that match the experimental spectrum (see Fig. 7.2(a)) and that have no chirp (corresponding to the gray curve in Fig. 7.2(b)) [200]. The layer structure of the sample is shown on the RHS. While propagating through the sample, the fundamental spectrum becomes significantly distorted, which leads to a lengthening of the pulse in time, and, thus to a reduction of the field amplitude and the Rabi frequency. This effect is most pronounced for pulses corresponding to the experiment (see Fig. 7.8(c)), where especially the sharp high-energy peak of the laser spectrum is largely affected. The effect is close to negligible for sech²-shaped pulses (Fig. 7.8(b)). The dispersive effects due to the linear dielectric function (Fig. 7.8(d)) further enhance the pulse distortions. Both effects lead to a reduction of the splitting of the Mollow sidebands around the third harmonic (compare Figs. 7.8(c) and 7.7). This explains the much (factor of two) smaller splitting seen in Fig. 7.3 as compared to the simple modeling (see white curves in Fig. 7.7). This reduced splitting between the Mollow sidebands obviously largely suppresses the anticipated interference term. Thus, the GaAs double heterostructure is not suitable for observing a dependence on the CEO phase. Also, it becomes obvious from Fig. 7.8 that the signal around the third harmonic varies very strongly with propagation coordinate z. This is mainly due to the fact that the absorption coefficients (for the third harmonic) of both, GaAs and $Al_{0.3}Ga_{0.7}As$, are on the order of $1/(10\,nm)$, which can easily be estimated from the corresponding linear dielectric functions shown in Fig. 7.8(d). As a dramatic result, the detected signal does not stem from the 600-nm thick GaAs layer sandwiched between $Al_{0.3}Ga_{0.7}As$ barriers – which was initially believed [181] – but rather from the thin GaAs cap layer initially employed as an antioxidation layer. The band edge of the $Al_{0.3}Ga_{0.7}As$ barriers leads to an (almost) off-resonant nonlinear signal that is expected from the dependence on detuning shown in Fig. 7.7.

Having noticed the relevance of propagation effects, it is interesting to see at what point the "intrinsic" response is recovered. Figure 7.9(a) shows the calculated spectra of light emitted into the forward direction of a thin layer of GaAs that has no $Al_{0.3}Ga_{0.7}As$ barriers for various values of the CEO phase ϕ. With such thin films, the meeting of the different Mollow sidebands can be observed. Furthermore, it becomes clear from the intensity dependence shown in Fig. 7.9(b) that it is *not* simply the interference of the tail of the laser spectrum itself (which is roughly equal to a box function) with the third-harmonic signal, but rather the interference of different Mollow sidebands. Figure 7.10 exhibits the same data as Fig. 7.9, but represented as

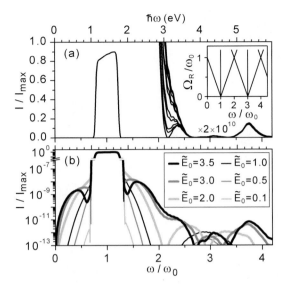

Fig. 7.9. (a) Signal intensity (linear scale, normalized to the maximum intensity, I_{max}, of the incident laser spectrum) emitted into the forward direction versus spectrometer frequency ω in units of the laser carrier frequency ω_0 for different values of the CEO phase ϕ. The GaAs film with $L = 20$ nm thickness on a substrate with $\epsilon_s = $ const. $= (1.76)^2$ has no $Al_{0.3}Ga_{0.7}As$ barriers on either side, but a front-side antireflection (AR) coating, $\tilde{E}_0 = 3.5 \times 10^9$ V/m, 5.6-fs $sinc^2$-shaped pulses. The inset illustrates the interference of the different Mollow sidebands as the Rabi frequency $\Omega_R = \hbar^{-1} d_{cv} \tilde{E}$ increases. **(b)** As (a), but signal intensity (normalized) on a logarithmic scale for fixed $\phi = 0$ and for different incident electric-field amplitudes (in units of 10^9 V/m) as indicated. Reprinted with permission from O. D. Mücke et al., Phys. Rev. Lett. **89**, 127401 (2002) [202]. Copyright (2002) by the American Physical Society.

a grayscale image. Obviously, the central peak of the third-harmonic Mollow triplet is much weaker than the outer two in such a rather realistic calculation.

Figure 7.11(a) depicts the intensity spectra of light emitted into the forward direction versus CEO phase ϕ for a $L = 100$-nm thin GaAs film on a substrate with $\epsilon_s = (1.76)^2$ (e.g., sapphire). Note the dependence on ϕ with large visibility around $\omega/\omega_0 = 2.05$ to 2.25 (this is a 284-meV or 38-nm broad interval) and the period of π according to (2.46) (rather than 2π according to (2.45)) resulting from the inversion symmetry of the problem. In other words: The signal does not depend on the sign of the electric field. (b) Same for $L = 20$ nm, indicating that one already has some distortions in (a) due to the finite thickness of the sample as a result of different group and phase velocities – the high-energy transitions do *not* react instantaneously as would be the case for the background dielectric constant. (c) As (b), but introducing a front-side $\lambda/4$-antireflection (AR) coating designed for the fundamental laser frequency ω_0. Note that (b) and (c) are shifted with respect to each other horizontally, because the incident optical pulses, and thus also ϕ, are distorted as a result of multiple reflections. (d) As (c), but for a different incident electric-field amplitude $\tilde{E}_0$. This variation also leads to a horizontal shift, which is both interesting as well

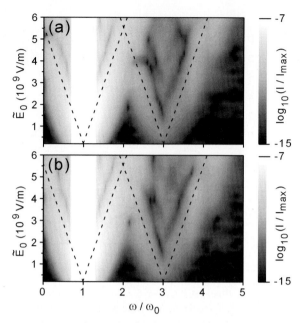

Fig. 7.10. Grayscale image of the emitted light intensity into the forward direction versus field amplitude of the incident pulses, corresponding to the inset in Fig. 7.9. Parameters as in Fig. 7.9(a). **(a)** $\phi = 0$, **(b)** $\phi = \pi/2$. Figure reprinted from Ref. [54] by permission of O. D. Mücke.

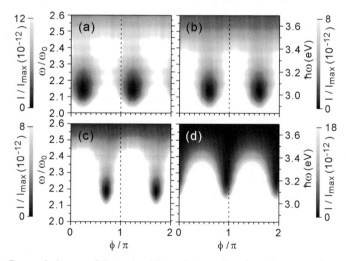

Fig. 7.11. Grayscale image of the emitted intensity as a function of ω and ϕ for a thin GaAs film with thickness L without $Al_{0.3}Ga_{0.7}As$ barriers on a substrate with dielectric constant ϵ_s. **(a)** $L = 100$ nm, $\tilde{E}_0 = 3.5 \times 10^9$ V/m, and $\epsilon_s = (1.76)^2$, **(b)** as (a) but $L = 20$ nm, **(c)** as (b) but with an additional front-side antireflection coating (as in Fig. 7.9), **(d)** as (c) but for an electric-field amplitude of $\tilde{E}_0 = 4.0 \times 10^9$ V/m. Reprinted with permission from O. D. Mücke et al., Phys. Rev. Lett. **89**, 127401 (2002) [202]. Copyright (2002) by the American Physical Society.

as disturbing. It is interesting on the one hand because no such intensity dependence occurs in off-resonant perturbative nonlinear optics. It is disturbing on the other hand, because in order to use the effect to determine the CEO phase, one needs to calibrate the incident electric-field amplitude, or, more precisely, the Rabi frequency. This is, however, possible in principle via the measured splitting of the Mollow sidebands.

7.1.3 Dependence on the Carrier-Envelope Phase

Let us come back to the experiment and connect to those sample designs just discussed in the theory. Figure 7.12 shows the spectra of light emitted into the forward direction for a $l = 100$-nm GaAs film epitaxially grown on a sapphire substrate. (a) corresponds to low excitation intensity, (b) to high excitation. The cuts through these data sets at

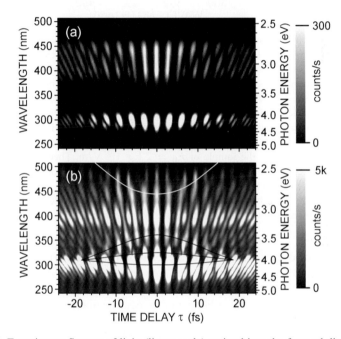

Fig. 7.12. Experiment: Spectra of light (linear scale) emitted into the forward direction by a $l = 100$-nm thin GaAs film on sapphire substrate resonantly excited by a pair of 5-fs pulses with time delay τ. The CEO phase ϕ of the laser pulses is not stabilized. **(a)** Excitation intensity $I = 0.24 \times 10^{12}$ W/cm^2, **(b)** $I = 2.8 \times 10^{12}$ W/cm^2 (both arms at $\tau = 0$). The contribution centered around $\lambda = 425$ nm wavelength is due to surface SHG. The single peak in (a) centered around $\lambda = 300$ nm wavelength (the third harmonic of the GaAs bandgap) evolves into three peaks in (b), which are attributed to the carrier-wave Mollow triplet. The corresponding three black lines are a guide to the eye. The white curve at the top of (b) (another guide to the eye) indicates the position of the high-energy peak of the fundamental Mollow triplet. For (b) we estimate that the peak Rabi energy *within* the GaAs film (and accounting for reflection losses at the air/GaAs interface) is given by $\Omega_R/\omega_0 = 0.76$. Figure reprinted from Ref. [54] by permission of O. D. Mücke.

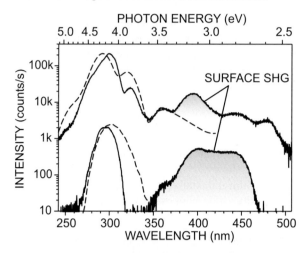

Fig. 7.13. Experiment (solid curves): Spectra of light (logarithmic scale) emitted into the forward direction by a $l = 100$-nm thin GaAs film on sapphire substrate excited by 5-fs pulses (cuts through Fig. 7.12 at $\tau = 0$). For low excitation intensity $I = 0.24 \times 10^{12}$ W/cm^2 (lower curve), well-separated peaks at the second harmonic and the third harmonic of the laser occur. When increasing the intensity to $I = 2.8 \times 10^{12}$ W/cm^2 (upper curve, shown on the same absolute scale), the deep valleys between the second harmonic and the third harmonic as well as between the fundamental and the second harmonic are filled and additional maxima are observed. We estimate that a "high" intensity of $I = 2.8 \times 10^{12}$ W/cm^2 translates into a peak electric field inside (!) the sample of $\tilde{E}_0 = 2.1 \times 10^9$ V/m. Theory (dashed curves): Corresponding spectra calculated from the semiconductor Bloch equations for "high" ($\tilde{E}_0 = 1.65 \times 10^9$ V/m) – also compare Fig. 7.16 – and "low" ($\tilde{E}_0 = 0.23 \times 10^9$ V/m) excitation, respectively. Reprinted with permission from Q. T. Vu et al., Phys. Rev. Lett. **92**, 217403 (2004) [182]. Copyright (2004) by the American Physical Society.

$\tau = 0$ are depicted in Fig. 7.13. At high excitation (Fig. 7.12(b)), the emitted light intensity around the third harmonic of the GaAs bandgap splits and overlaps with the second-harmonic generation (SHG) signal. From the dependence on l (not shown) we conclude that the SHG has a large surface contribution (or is even completely generated at the two GaAs surfaces), while the third harmonic is consistent with a bulk effect. At $\tau = 0$, the spectrum exhibits three peaks around the third harmonic that evolve with time delay τ. The solid lines are guides to the eye and indicate that the splitting decreases with increasing $|\tau|$. These three peaks are interpreted as the carrier-wave Mollow triplet. Note also that a contribution from the fundamental moves into the picture from the top. Following the above theory, this is expected to be the high-energy peak of the fundamental Mollow triplet. The data of the $l = 50$ nm sample (not shown) are compatible with the $l = 100$ nm sample data (Fig. 7.12(a) and (b)), however – as already discussed above – the second-harmonic contribution is more prominent with respect to the third harmonic in the $l = 50$ nm case as compared to the $l = 100$ nm case due to a larger surface contribution. This significantly reduces the visibility of the low-energy peak of the third-harmonic Mollow triplet.

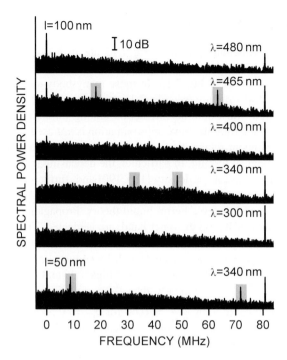

Fig. 7.14. Experiment: Radio-frequency power spectra (logarithmic scale), 10 kHz resolution and video bandwidth, for various optical detection wavelengths λ and two GaAs film thicknesses l as indicated. The spectra are vertically displaced for clarity. The excitation intensity is comparable to that in Fig. 7.12(b), $\tau = 0$. The peaks at the CEO frequency f_ϕ and at $(f_r - f_\phi)$ are highlighted by gray areas. Figure reprinted from Ref. [54] by permission of O. D. Mücke.

Fig. 7.14 shows measured radio-frequency (RF) power spectra (also see Sect. 2.6) of the signals corresponding to Fig. 7.12(b), $\tau = 0$ for 20-nm broad wavelength intervals centered around the quoted center wavelength. The peak in the RF spectra at 81 MHz arises from the repetition frequency f_r of the laser oscillator. From the optical spectra depicted in Figs. 7.12 and 7.13 one expects an optimum interference of the high-energy fundamental Mollow triplet with the surface SHG around $\lambda = 465$ nm and an optimum interference of the surface SHG with the low-energy third-harmonic Mollow triplet around $\lambda = 340$ nm. At these wavelengths, Fig. 7.14 does indeed show peaks at the CEO frequency f_ϕ and at difference frequency $(f_r - f_\phi)$. The value of f_ϕ changes from measurement to measurement because the intracavity prism near the high-reflector is moved intentionally to demonstrate the influence of intracavity dispersion on the results. For other detection wavelengths shown in Fig. 7.14, no corresponding peaks are observed, even though the absolute signal levels are larger (see larger f_r peak). Note that the f_ϕ and $(f_r - f_\phi)$ peaks in the RF power spectrum are less than 8 dB smaller than the f_r peak, indicating that the relative modulation depth of the beat signal versus time is as large as 40%. Similar results are observed for the $l = 50$-nm thin sample (see lowest data set in Fig. 7.14).

7.1.4 Semiconductor Bloch Equations

The measured experimental spectra in Fig. 7.13 are in good agreement with microscopic calculations [182] on the basis of the semiconductor Bloch equations, depicted as dashed curves in Fig. 7.13. In these many-body-system calculations, a full tight-binding band structure (see Fig. 7.1) rather than the effective-mass approximation is employed, the Coulomb interaction among the carriers is treated on a Hartree–Fock level, the rotating-wave approximation is *not* used and the energy- and carrier-density-dependent relaxation and dephasing processes are accounted for on a semiphenomenological footing. In particular, the dephasing rate roughly increases with the third root of the carrier density [186, 190] and, furthermore, increases with increasing electron momentum because of a much larger final density of states – qualitatively similar to Landau's Fermi liquid theory. Propagation effects as well as the (extrinsic) surface SHG, on the other hand, are neglected. Thus, experiment and theory can only be compared directly around the third harmonic. Furthermore, intraband processes (see Sect. 4.2) are neglected. Indeed, at the peak electric field of $\tilde{E}_0 = 1.65 \times 10^9$ V/m used in the calculations of Figs. 7.13 and 7.16, the electron ponderomotive energy is $\langle E_{kin} \rangle = 0.37\,e$V, thus $\frac{\langle E_{kin} \rangle}{\hbar \omega_0} = 0.27$ (compare Example 4.1), which has to be compared with $\frac{\hbar \Omega_R}{\hbar \omega_0} = 0.60$ for $d_{cv} = 0.5\,e$ nm. This comparison might justify this approximation. From the agreement between experiment and theory in Fig. 7.13 – also note the absolute values of the peak electric field $\tilde{E}_0$ – we conclude that our above simple and intuitive discussion of the peaks around the third harmonic in terms of the carrier-wave Mollow triplet remains correct in the presence of energy bands (rather than discrete energy levels) in a semiconductor. The center of this triplet is redshifted with respect to the photon energy $\hbar \omega = 3E_g = 4.26\,e$V because of renormalization effects.

An intuitive understanding for the case of band-to-band transitions can be obtained for continuous-wave excitation, which is illustrated in Fig. 7.15. Figure 7.15(a) again shows the (fundamental) two-level system Mollow triplet. For excitation within the bands of a semiconductor it has to be replaced by the two-band Mollow triplet arising from light-induced gaps [203–206]. These gaps are highlighted by the gray areas in Fig. 7.15(b). The light-induced gap in, e.g., the conduction band arises because the original conduction band and the one-photon sideband of the valence band lead to an avoided crossing. The corresponding Hopfield coefficients [182] determine the amount of conduction band admixture. It is crucial to note that this admixture can be finite even far away from the fictitious crossing point. This statement becomes particularly important for Rabi energies $\hbar \Omega_R$ approaching the photon energy $\hbar \omega_0$, while it can be neglected for small Rabi energies. Importantly, as a result of this finite admixture, optical transitions between the induced bands are possible throughout an appreciable fraction of momentum space. Consequently, the transitions acquire considerable spectral weight. Two out of the four sets of possible optical transitions are energetically degenerate and a triplet results. The optical transitions corresponding to the low-energy peak of the two-band Mollow triplet in Fig. 7.15 lead to induced absorption well below the original bandgap energy, which is qualitatively consistent with the experimental observations shown in Fig. 7.6.

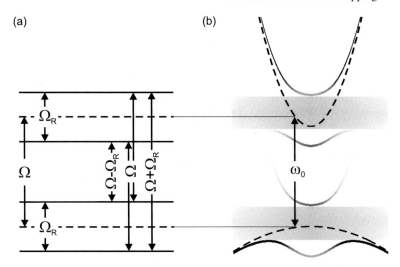

Fig. 7.15. (a) Scheme of the two-level system Mollow triplet for resonant excitation, i.e., $\Omega = \omega_0$. Also compare with Fig. 3.15. **(b)** Light-induced gaps (gray areas) are the analogue for interband transitions in a semiconductor. The square modulus of the Hopfield coefficients of the four resulting bands (frequency versus electron wave number) is superimposed onto them by a grayscale: Black corresponds to 1, white to 0. GaAs parameters are used, the Rabi energy is given by $\hbar\Omega_R/E_g = 0.5$, and $\hbar\omega_0/E_g = 1.1$. The dashed curves show the original conduction and valence band, respectively. Both (a) and (b) assume monochromatic excitation as well as negligible damping and are based on the rotating-wave approximation. Reprinted with permission from Q. T. Vu et al., Phys. Rev. Lett. **92**, 217403 (2004) [182]. Copyright (2004) by the American Physical Society. It is interesting to compare (b) with the case of intraband excitation shown in Fig. 4.2 (note the inverted grayscale).

Figure 7.16 shows the calculated inversion versus time for various optical excess energies with respect to the unrenormalized bandgap E_g (as indicated). Parameters are identical to "high" excitation in Fig. 7.13. Structures with twice the frequency of light due to the carrier-wave oscillation are clearly visible. As discussed above, these structures are intimately related to the third-harmonic generation. At small excess energies (close to the band edge), the inversion nearly performs a complete Rabi flop during the first half of the pulse ($t < 0$). A second Rabi flop roughly between $t = -1$ fs and $t = 6$ fs is already very heavily damped. At later times, dissipative rather than coherent kinetics dominates and the low-energy states are filled from above. At large excess energies, where the dephasing is stronger, only an irreversible increase of the inversion is left after the pulse has passed. Note that the computed electron–hole pair density after the pulse is as large as $n_{eh} = 1.1 \times 10^{20}$ /cm^3 – the material changes from a semiconductor at $t \approx -7$ fs in Fig. 7.16 into a metal at $t \approx -4$ fs within just one cycle of light.

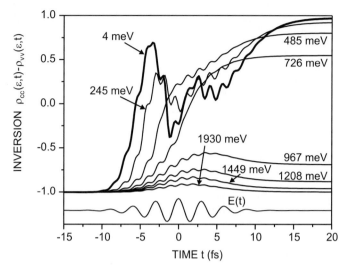

Fig. 7.16. Semiconductor Bloch equations: Inversion for various excess energies ϵ (as indicated) versus time t for a peak electric field of $\tilde{E}_0 = 1.65 \times 10^9$ V/m. The carrier photon energy of the 7-fs pulses with Gaussian envelope is $\hbar\omega_0 = 1.38\,eV$, the CEO phase is $\phi = 0$. The resulting laser electric field $E(t)$ is illustrated in the lower part. Compare with Fig. 7.13. For $t = 20\,\mathrm{fs}$, the carrier density is as large as $n_{\mathrm{eh}} = 1.1 \times 10^{20}\,/\mathrm{cm}^3$. Reprinted with permission from Q. T. Vu et al., Phys. Rev. Lett. **92**, 217403 (2004) [182]. Copyright (2004) by the American Physical Society.

7.2 "THG in the Disguise of SHG"

Another striking effect of extreme nonlinear optics is the occurrence of frequency doubling from two-level systems. This phenomenon can even occur in a medium with a center of inversion. We have seen in Sect. 3.4 that it is most pronounced for a transition frequency Ω around twice the carrier frequency of light, i.e., for $\Omega/\omega_0 = 2$. For a carrier photon energy of $\hbar\omega_0 = 1.5\,eV$, we are looking for a transition energy around $3\,eV$. In a medium without inversion symmetry, this mechanism of "THG in the disguise of SHG" adds to the usual SHG via a $\chi^{(2)}$ process.

Samples

The direct gap semiconductor ZnO has a room temperature bandgap energy of $E_{\mathrm{g}} = 3.3\,eV$. Amazingly, the precise value of the bandgap energy of ZnO was subject of scientific discussions until rather recently [207]. ZnO has a *c*-axis without inversion symmetry and is birefringent ($\boldsymbol{E}||\boldsymbol{c}$ and $\boldsymbol{E} \perp \boldsymbol{c}$ are inequivalent). For ZnO single-crystal platelets (here about $100\,\mu\mathrm{m}$ thick), the *c*-axis lies within the plane of the platelet, while for the 350-nm thin ZnO epitaxial film discussed in this book, the *c*-axis is perpendicular to the film. The ZnO interband dipole matrix element is smaller than for GaAs. From $\boldsymbol{k} \cdot \boldsymbol{p}$ perturbation theory according to (7.1) we obtain the ZnO dipole matrix element of $d_{\mathrm{cv}} = 0.19\,e\,\mathrm{nm}$.

Experiment

In the ZnO experiments [208] the same 5-fs laser system with $\hbar\omega_0 \approx 1.5\,eV$, the same stabilized Michelson interferometer and the same focusing optics as in the GaAs experiments are used. Also, the reference intensity $I_0 = 0.6 \times 10^{12}\,\mathrm{W/cm^2}$ (one arm of the interferometer) has the same definition and the same value. The light emitted by the samples into the forward direction is spectrally filtered (3 mm *Schott BG 39*) to remove the prominent fundamental laser spectrum and is sent into a 0.25-m focal length grating spectrometer connected to a charge-coupled-device (CCD) camera. Alternatively, the light is sent through a combination of filters (3 mm *Schott BG 39*, 3 mm *Schott GG 455*, and *Coherent 35-5263-000* 480 nm cutoff interference filter for the 100-μm ZnO single crystal and *Coherent 35-5289-000* 500 nm cutoff interference filter for the 350-nm epitaxial film, respectively) onto a 50-Ω-terminated photomultiplier tube (*Hamamatsu R 4332, Bialkali photocathode*), connected to an RF spectrum analyzer (*Agilent PSA E4440A*).

Figure 7.17 shows measured optical spectra in the spectral region energetically above the laser spectrum (which has its short wavelength cutoff above 650 nm) and below the bandgap of ZnO of $E_g = 3.3\,eV$ for (a) low excitation and (b), (c) large excitation intensity versus time delay τ of the interferometer. Note that (b) and (c) are the same data plotted with different levels of saturation in order to reveal details. The measured intensity is given in actual counts per second (one count corresponds to about two photons). All spectral components shown in Fig. 7.17 are also easily visible with the naked eye. If the ZnO sample is moved out of the focus by some tens of micrometers, all the spectral components shown in Fig. 7.17 completely disappear, indicating that none of them comes directly out of the laser. Polarization-dependent experiments under these conditions show that all these spectral components have the same linear polarization as the laser pulses. In (a), the light around 390–470 nm wavelength is due to second-harmonic generation (SHG), the components above 500 nm are due to self-phase modulation (SPM). Interestingly, the independently measured interferometric autocorrelation of the laser pulses (using a thin beta barium borate (BBO) SHG crystal, see curve labeled IAC in Fig. 7.17(b)), is closely reproduced by a cut at 395 nm wavelength (see white curve in Fig. 7.17(c)). This indicates that the pulses are not severely broadened in the ZnO crystal due to, e.g., group velocity dispersion. Furthermore, the spectral width of the SHG contribution indicates that phase-matching effects do not play a major role, which is not surprising considering the short Rayleigh range of the microscope objective of only several micrometers (see Fig. 7.5(a)). For higher intensities, the spectral overlap of SPM and SHG becomes immediately obvious from the spectra in Fig. 7.17(c) and a rich fine structure as a function of τ appears in this spectral region. Feeding the spectral components of this interference region into an RF spectrum analyzer, we find clear evidence for a peak at the carrier-envelope offset frequency f_ϕ (Fig. 7.18) that arises because ϕ changes from pulse to pulse of the mode-locked laser oscillator due to different group delay and phase delay times per round trip of the laser cavity (see Sect. 2.3). To further check this assignment, we also depict the RF spectrum for a slightly different laser end mirror position, which shifts the f_ϕ peak as well as the mixing product $(f_r - f_\phi)$

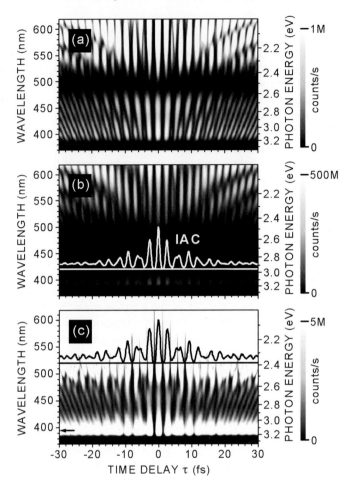

Fig. 7.17. Experiment: Spectra of light emitted by the 100-μm thick ZnO single crystal into the forward direction versus time delay τ of the Michelson interferometer, $E||c$. **(a)** $I = 0.15 \times I_0$, **(b)** $I = 2.04 \times I_0$, and **(c)** as (b), but different saturation of the grayscale. The light intensities decay near the ZnO bandgap of $E_g = 3.3\,eV$. The white curve in (b) labeled IAC is the independently measured interferometric autocorrelation using a BBO crystal, the black curve in (c) is a cut through the ZnO data at 395 nm wavelength (see arrow). Reprinted with permission from O. D. Mücke et al., Opt. Lett. **27**, 2127 (2002) [208]. Copyright (2002) by the Optical Society of America.

(see labels in Fig. 7.18). In the 350-nm thin ZnO film (Fig. 7.18(b)), both, the f_ϕ peak as well as the $2f_\phi$ peak are still visible. Interestingly, the $2f_\phi$ peak is even larger than the f_ϕ peak. The fact that the relative height of the $2f_\phi$ peak is much smaller for the ZnO single crystal (which is much thicker than the Rayleigh length, see Fig. 7.5(a)) than for the 350-nm film, is likely due to the variation of the CEO phase along the propagation direction arising from the Gouy phase (see Sect. 6.3). This effect, which

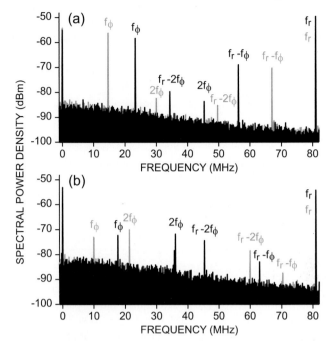

Fig. 7.18. Experiment: RF spectra, 10 kHz resolution and video bandwidth, $\tau = 0\,\text{fs}$ (equivalent to an average total power of 64 mW in front of the sample). **(a)** 100-μm thick ZnO single crystal, $\boldsymbol{E}\|\boldsymbol{c}$, corresponding to Figs. 7.17 (b), (c), optical filter roughly corresponds to 455–480 nm, **(b)** 350-nm thin ZnO epitaxial layer, $\boldsymbol{E} \perp \boldsymbol{c}$, optical filter roughly corresponds to 455–500 nm. The peaks at the repetition frequency f_r, the carrier-envelope offset frequency f_ϕ, its second harmonic $2f_\phi$ and the mixing products $(f_r - f_\phi)$ and $(f_r - 2f_\phi)$ are labeled. The black and gray data correspond to slightly different laser end mirror positions. When removing intracavity prism material (CaF$_2$), the f_ϕ and $2f_\phi$ ($(f_r - f_\phi)$ and $(f_r - 2f_\phi)$) peaks shift to the left (right). Reprinted with permission from O. D. Mücke et al., Opt. Lett. **27**, 2127 (2002) [208]. Copyright (2002) by the Optical Society of America.

clearly affects the $2f_\phi$ peak more strongly than the f_ϕ peak, is negligible for the 350-nm thin ZnO film.

If one is interested not only in determining a dependence on the CEO phase ϕ but rather interested in measuring ϕ of the incident pulse itself, it is obviously relevant to ask whether ϕ changes while propagating through the sample (as we have done for the GaAs case in Sect. 7.1.2). Apart from the Gouy phase just mentioned, two effects can change the CEO phase within the sample: Linear optical propagation effects and nonlinear optical effects. The magnitude of the linear optical propagation effects is much easier to estimate for the ZnO case as compared to the GaAs case because of the off-resonant excitation conditions. In other words: Absorption plays no role here. Under these conditions, for a carrier frequency ω_0 and a center vacuum wavelength

$\lambda_0 = 2\pi c_0/\omega_0$, the change of ϕ of the pulse, $\delta\phi$, as a result of propagation over a length l results from the different group and phase velocities according to

$$\delta\phi = 2\pi \frac{\left(\dfrac{l}{v_{\text{group}}} - \dfrac{l}{v_{\text{phase}}}\right)}{2\pi/\omega_0} \tag{7.17}$$

$$= -2\pi \frac{dn}{d\lambda}(\lambda_0)\, l\,,$$

with the vacuum-wavelength-dependent refractive index $n(\lambda)$. From the first to the second line in (7.17) a few straightforward mathematical manipulations are necessary. For many materials, the dependence $n(\lambda)$ is parametrized using the so-called Sellmeier formula[1] according to

$$n^2(\tilde{\lambda}) = \epsilon(\tilde{\lambda}) = \mathcal{A} + \frac{\mathcal{B}\tilde{\lambda}^2}{\tilde{\lambda}^2 - \mathcal{C}^2} + \frac{\mathcal{D}\tilde{\lambda}^2}{\tilde{\lambda}^2 - \mathcal{E}^2} \tag{7.18}$$

with $\tilde{\lambda} = \lambda/(0.1\,\text{nm})$. For the relevant wavelengths λ, the refractive index according to (7.18) is real. Sun and Kwok [209] determined the fit parameters $\mathcal{A} = 2.0065$, $\mathcal{B} = 1.5748 \times 10^6$, $\mathcal{C} = 10^8$, $\mathcal{D} = 1.5868$, and $\mathcal{E} = 2606.3$ for ZnO, $\boldsymbol{E} \perp \boldsymbol{c}$. This fit is applicable to the visible part of the spectrum only. With (7.17) for $\lambda_0 = 826\,\text{nm}$ ($\Leftrightarrow \hbar\omega_0 = 1.5\,e\text{V}$) and after some tedious mathematics, this finally delivers

$$\delta\phi_{\text{ZnO},\,\lambda_0=826\,\text{nm}} = 0.013 \times 2\pi\, l/100\,\text{nm}\,. \tag{7.19}$$

For the above experiments with 100-μm thick ZnO single crystals, the effective interaction length l is given by the depth of focus (see Fig. 7.5(a)) that is on the order of five micrometers, in which case we have $\delta\phi = 0.7 \times 2\pi$ – a significant change of the CEO phase within the sample. For the $l = 350\,\text{nm}$ thin ZnO epitaxial film, $\delta\phi$ is merely 4.6% of 2π, which might be sufficiently small for many applications.

Theory

From our discussion in Sect. 2.6 it is clear that the peaks in the RF spectra at f_ϕ originate from an interference of the fundamental with the second harmonic, while the peaks at $2f_\phi$ and the mixing products ($f_r - 2f_\phi$) in Fig. 7.18 stem from an interference of the fundamental and the third harmonic. *This is really amazing!* The centers of the fundamental and the third harmonic are separated by twice the laser carrier frequency – yet, they do interfere. Is a simple description of the nonlinear optical polarization in terms of nonlinear optical susceptibilities according to (2.37) going to work at all? To see how it works, we solve (2.37) together with the one-

---

[1] This is simply the dielectric function of the sum of two resonances as, e.g., from the optical Bloch equations, plus a constant.

dimensional Maxwell equations (see Sect. 6.1) numerically, i.e., we do not employ the slowly varying envelope approximation and work with the real electric laser field $E(z, t)$ [200] and with the actual layer structure of the sample, i.e., a $l = 350$ nm thin layer of ZnO on a semi-infinite sapphire substrate. The latter has a dielectric constant of $\epsilon_s = (1.76)^2$ (compare Sect. 7.1.2, i.e., its optical nonlinearities are neglected[2]. For the description of the ZnO layer we use the nonlinear optical susceptibilities $\chi^{(2)}$ and $\chi^{(3)}$ [67]. Terms of order four and higher are neglected. On this level of modeling, the optical spectra of the experiment are nicely reproduced (not shown). Inspection of the corresponding calculated RF power spectrum (calculated on the basis of (2.50) and (2.52)) shows a peak at frequency f_ϕ but no contribution at frequency $2f_\phi$ (not shown), while in the experiment the $2f_\phi$ peak is even larger than the f_ϕ peak (see Fig. 7.18(b)). Thus, we conclude that *the $2f_\phi$ peak in the RF power spectra of ZnO can definitely not be explained on the basis of perturbative off-resonant nonlinear optics*.

The solution [67] to the "riddle of the $2f_\phi$ peak" has already been visible in Figs. 3.7 and 3.8. The signal contribution around the $\Omega = \omega$ line in Fig. 3.7 exhibits a constriction, the exact position of which depends on the Rabi frequency. Let us start the discussion with the part above this constriction and consider a transition frequency Ω on the vertical axis at the second harmonic of the laser carrier frequency ω_0, i.e., at $\Omega/\omega_0 = 2$ in Fig. 3.7(a). Here we observe a well-defined peak in the optical spectra right at the spectrometer frequency $\omega = 2\omega_0$ (see white curve in Fig. 3.7(a) that is a cut through the data at $\Omega/\omega_0 = 2$ plotted on a *linear* scale). What is the origin of this peak? A part of it is the resonant enhancement way down in the low-energy tail of the third harmonic of the laser photon energy. For pulses containing many cycles of light, this contribution would disappear because of negligible overlap of the third-harmonic response function and the resonance – it is specific for the regime of few-cycle pulses (see dependence on pulse duration depicted in Fig. 3.7(b)). This third-harmonic contribution is expected to be associated with a phase 3ϕ – even though it peaks right at the spectrometer frequency $\omega = 2\omega_0$. Its signal strength roughly scales with the third power of the laser intensity. The part of the $\Omega = \omega$ signal contribution below this constriction can be interpreted as the resonantly enhanced SPM due to absorption of photons from the high-energy tail of the laser spectrum with phase 1ϕ. Its signal strength is roughly proportional to the intensity itself. Thus, the upper part gains relative weight with respect to the lower part for increasing intensity or increasing Rabi energy Ω_R in Figs. 3.6 and 3.8.

Figure 3.7(c) shows the dependence on the CEO phase ϕ for a selected transition frequency of $\Omega/\omega_0 = 2$. All other parameters are as in Fig. 3.7(a). It becomes obvious that a part of the interference occurs in between the fundamental, i.e., $\omega/\omega_0 = 1$, and $\omega/\omega_0 = 2$. Note that the period of the signal versus ϕ is π rather than 2π, equivalent to a peak at frequency $2f_\phi$ in the RF spectrum. The usual SHG would appear in the same region in the optical spectra as the *third harmonic in the disguise of a second harmonic*, but its phase is 2ϕ rather than 3ϕ, thus, it leads to a peak at frequency f_ϕ in

---

[2] Indeed, no measurable nonlinear signal of the sapphire substrate itself occurs in independent additional experiments.

the RF spectra when beating with the fundamental. Another part of the interference in Fig. 3.7(c) occurs in between $\omega/\omega_0 = 2$ and $\omega/\omega_0 = 3$. This shows that the peak around $\omega/\omega_0 = 2$ in Fig. 3.7(a) is indeed a mixture of resonantly enhanced SPM and resonantly enhanced THG.

Figure 3.6 illustrates the dependence on Rabi energy. For large Rabi energies (d), the *"THG in the disguise of SHG"* becomes the dominating feature in the optical spectrum (black curve at $\Omega/\omega_0 = 2$). The unusually small contribution of P around the laser carrier frequency, i.e., at $\omega/\omega_0 = 1$ in Fig. 3.6(d) is due to carrier-wave Rabi flopping (see Sect. 3.3). Indeed, the inversion w starts at -1, reaches values near $+1$ in the maximum and comes almost back to -1 after the pulse – even though the excitation is "off-resonant" with $\Omega/\omega_0 = 2$. This illustrates the fact that the detuning of the carrier frequency from resonance becomes negligible if the Rabi energy is larger than the detuning.

If one interprets the transition energy $\hbar\Omega$ in Fig. 3.7(a) as the bandgap energy E_g of a semiconductor, the lower RHS triangle formed by the $\Omega = \omega$ line experiences strong reabsorption in the semiconductor band-to-band continuum, while the upper LHS triangle is in the transparency regime of the semiconductor. *"THG in the disguise of SHG"* overlaps with this line. In order to study corresponding reabsorption and phase-matching effects, we now present numerical solutions of the coupled Maxwell–Bloch equations in one dimension without using the rotating-wave approximation and without using the slowly varying envelope approximation and accounting for the actual sample geometry, i.e., we model a 350-nm thin film of ZnO on a sapphire substrate with dielectric constant $\epsilon_s = (1.76)^2$ using a one-dimensional finite-difference time-domain algorithm. Furthermore, for a semiconductor, one does not have a single two-level system but rather a band continuum, i.e., one needs to integrate P along the vertical axis in Fig. 3.7(a). To be close to the experiment, we fit an ensemble of 45 two-level systems (3.3–7.9 eV) to the known measured linear dielectric function of ZnO over a broad frequency regime [210] (see Fig. 7.19). However, the optical transitions at large photon energies do not contribute to the signal at the spectrometer frequency $\omega/\omega_0 = 2$ (for example, see $\hbar\Omega = 4$–$6\,eV$ on the RHS vertical axis of Fig. 3.7(a), where we have chosen $\hbar\omega_0 = 1.5\,eV$). On the other hand, they do contribute to THG and SPM – they essentially still act as an off-resonant $\chi^{(3)}$ susceptibility under these conditions. We have seen that a $\chi^{(3)}$ susceptibility (together with a $\chi^{(2)}$ susceptibility) is unable to reproduce the experimental data. We conclude that the high-energy transitions can only have a minor contribution to the nonlinear optical response. This is consistent with our findings for the GaAs case (see Sect. 7.1.2), where also predominantly transitions close the bandgap contributed to the nonlinear optical response (as a result of Coulomb correlations and much stronger damping/relaxation of high-energy states). Consequently, in the ZnO modeling, optical transitions from the bandgap at $E_g = 3.3\,eV$ up to $0.1\,eV$ above the gap are treated via the complete Bloch equations (using the ZnO dipole matrix element of $d_{cv} = 0.19\,e\,nm$, $T_1 = \infty$, and $T_2 = 34\,fs$), whereas the nonlinearity of the transitions at higher energies is suppressed by setting their inversion in the Bloch equations to -1 (as in the GaAs case). In this fashion, linear optical propagation effects are still

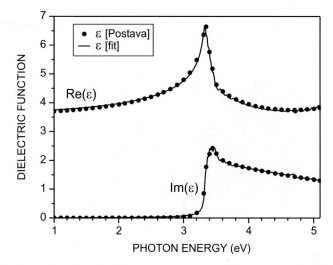

Fig. 7.19. Dielectric function of ZnO versus (spectrometer) photon energy $\hbar\omega$. The dots represent the data measured in Ref. [210], the full curve is a "fit" using an ensemble of 45 two-level systems with different transition frequencies Ω. This ensemble is used for the calculations presented in Fig. 7.20 [269].

accounted for exactly. In addition to this, ZnO has no inversion symmetry and shows a nonzero $\chi^{(2)}$ susceptibility. For simplicity, we describe this aspect by an effective bulk frequency-independent $\chi^{(2)} = 1 \times 10^{-13}$ m/V (remember that we have $\boldsymbol{E} \perp \boldsymbol{c}$ for the 350-nm ZnO film and that surface SHG can become important for such thin films). Furthermore, we employ excitation with a pair of collinearly propagating identical pulses with time delay τ. The pulses are taken directly from the experiment [200] (see Sect. 7.1.2). The ratio of the pulse repetition frequency f_r and the CEO frequency f_ϕ is set to $f_r/f_\phi = 5$, with $f_r = 81$ MHz.

In Fig. 7.20(a) we depict the calculated optical spectra versus time delay τ. Note that none of the spectral components originate from the incident pulses directly, all of them are rather generated in the 350-nm thin ZnO layer. The spectral components above about 520 nm wavelength are predominantly due to SPM, those in the range from 365 nm to 455 nm wavelength are mainly due to a combination of conventional SHG and *"THG in the disguise of SHG"*. In between the two regions, interference leads to a dependence on the CEO phase ϕ. Indeed, filtering out this region and computing the corresponding RF spectrum delivers peaks at the CEO frequency f_ϕ and at 2 f_ϕ (Fig. 7.20(b)) as expected from our above reasoning.

Figure 7.21 shows the experimental results corresponding to Fig. 7.20. The agreement is good, both for the optical spectra as well as for the RF spectra. In particular, note that the heights of the peaks at frequencies f_ϕ and 2 f_ϕ in the RF spectra are comparable, both for theory and experiment, indicating that *"THG in the disguise of SHG"* is comparable in magnitude to conventional SHG under these conditions. Furthermore, the nodal lines in the optical spectra indicated by the thin white lines

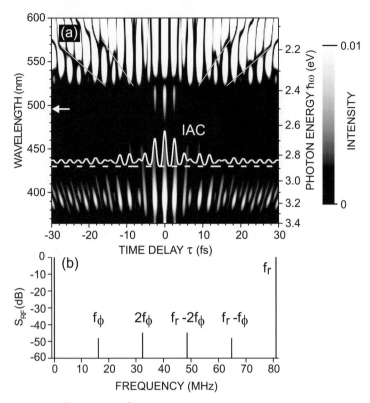

Fig. 7.20. Theory, $\tilde{E}_0 = 8 \times 10^9\,\mathrm{V/m}$ (in air) at $\tau = 0$. **(a)** Grayscale image of the light intensity (normalized) emitted into the forward direction versus spectrometer photon energy $\hbar\omega$ (linear wavelength scale) and time delay τ. The data are averaged over $\phi = 0$ to 2π. The thin white lines are a guide to the eye. The white curve labeled IAC is the interferometric autocorrelation of the laser pulses. **(b)** Radio-frequency power spectrum S_{RF} of the intensity at the spectral position indicated by the arrow on the LHS in (a), $\tau = 0$ [269].

in Fig. 7.21(a) qualitatively match those obtained for the complete modeling (see Fig. 7.20(a)). These nodal lines, however, are closely similar if the nonlinear response is described by an off-resonant perturbative $\chi^{(3)}$ process (not shown). In contrast, the narrow peak centered around 520 nm wavelength near $\tau = 0$ (where the peak intensity from the interferometer is maximum) does not occur (not shown) if the Bloch equations are replaced by an off-resonant perturbative $\chi^{(3)}$ process. Thus, this spectral peak, which appears in both experiment (Fig. 7.21(a)) and theory (Fig. 7.20(a)), is another signature of extreme nonlinear optics under these conditions.

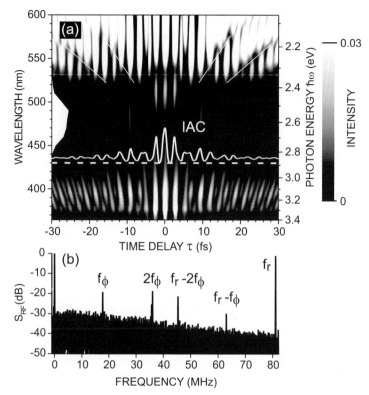

Fig. 7.21. Experiment on the 350-nm ZnO film, $\tilde{E}_0 = 6 \times 10^9$ V/m (in air) at $\tau = 0$. **(a)** Grayscale image of the light intensity (normalized) emitted into the forward direction versus spectrometer photon energy $\hbar\omega$ (linear wavelength scale) and time delay τ. ϕ is not stabilized. The thin white lines are a guide to the eye. The white curve labeled IAC is the interferometric autocorrelation of the laser pulses, obtained from an independent measurement using a β-barium borate SHG crystal. **(b)** Radio-frequency power spectrum S_{RF} of the intensity in the spectral interval indicated by the gray area on the LHS in (a), $\tau = 0$. Note the good agreement with the theory calculated under the same conditions (Fig. 7.20). Reprinted with permission from T. Tritschler et al., Phys. Rev. Lett. **90**, 217404 (2003) [67]. Copyright (2003) by the American Physical Society.

7.3 Dynamic Franz–Keldysh Effect

Let us first briefly remind ourselves about the normal static Franz–Keldysh effect and then come to the dynamic Franz–Keldysh effect, another example of extreme nonlinear optics, which was observed in experiments rather recently.

Static-field aproximation

The main [212] theoretical prediction by Franz [211] and Keldysh in 1958 was that applying a static electric field on the order of 10^7 to 10^8 V/m to a bulk direct-gap semiconductor, results in an exponential absorption tail below the fundamental bandgap. Later, additional spectral features above the bandgap were discovered. All these features can be understood in terms of photon-assisted tunneling of electrons from the valence to the conduction band: In the presence of the static electric field $\tilde{E}_0 > 0$ along the x-direction, the stationary one-dimensional Schrödinger equation becomes

$$-\frac{\hbar^2}{2m_e}\frac{\partial^2}{\partial x^2}\psi(x) + x\,e\,\tilde{E}_0\,\psi(x) = E_e\,\psi(x), \tag{7.20}$$

with the electron energy E_e. Substituting

$$X = x\left(\frac{2m_e e\tilde{E}_0}{\hbar^2}\right)^{1/3} - \frac{2m_e E_e}{\hbar^2}\left(\frac{2m_e e\tilde{E}_0}{\hbar^2}\right)^{-2/3} \tag{7.21}$$

leads to the standard (nonlinear) differential equation

$$\frac{d^2\psi}{dX^2}(X) = X\,\psi(X). \tag{7.22}$$

Its solution is $\psi(X) = \mathrm{Ai}(X)$ with the Airy function $\mathrm{Ai}(X)$, which decays exponentially for $X \to +\infty$ and that oscillates with a wavelength proportional to $1/\sqrt{|X|}$ for negative X. Additional properties of the Airy function can be found in mathematical handbooks [80]. The important aspect for our purposes is that this wave function has a tail that extends into the gap of the semiconductor. This is illustrated in Fig. 7.22. For an electron wave function and a hole wave function separated in energy by less than the gap energy E_g, the probability to find and electron and a hole at the same position has become finite. The larger the applied electric field, the larger the overlap. This allows for optical transitions at photon energies $\hbar\omega < E_g$ below the gap energy E_g. The resulting shape of the absorption spectrum $\alpha(\hbar\omega)$ within the effective-mass approximation is given by [199]

$$\alpha(\hbar\omega) \propto |\tilde{E}_0|^{1/3}\left(\tilde{\omega}\left(\mathrm{Ai}(-\tilde{\omega})\right)^2 + \left(\mathrm{Ai}'(-\tilde{\omega})\right)^2\right), \tag{7.23}$$

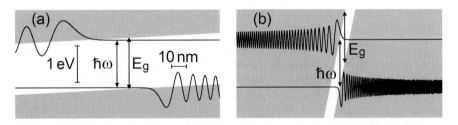

Fig. 7.22. Illustration of the electron and hole ($m_e \to m_h$) wave function for an optical transition at photon energy $\hbar\omega$ in a semiconductor subject to a static electric field $\tilde{E}_0$. Room temperature GaAs parameters are used. $E_g = 1.42\,eV$, $\hbar\omega = 1.3\,eV$. **(a)** $\tilde{E}_0 = 2 \times 10^6$ V/m, **(b)** $\tilde{E}_0 = 2 \times 10^8$ V/m (same scale). The gray areas represent the tilted bands [268].

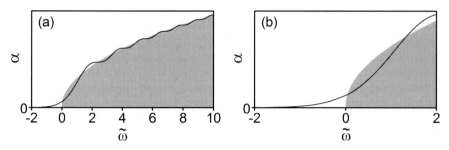

Fig. 7.23. Absorption spectra of a three-dimensional semiconductor subject to a static electric field according to (7.23) versus normalized spectrometer frequency $\tilde{\omega}$ according to (7.24). For example, for $\tilde{E}_0 = 10^8$ V/m (equivalent to an intensity of $I = 1.3 \times 10^9$ W/cm^2) and GaAs parameters, i.e. exciton binding energy $E_b = 4\,meV$ and exciton Bohr radius $r_B = 20$ nm, a change of $\tilde{\omega}$ equal to 1 corresponds to a change in the actual photon energy $\hbar\omega$ of 0.25 eV. $\tilde{\omega} = 0 \Leftrightarrow \hbar\omega = E_g$. The gray areas illustrate the zero-field absorption spectrum according to (7.25). **(a)** and **(b)** show the universal behavior $\alpha(\tilde{\omega})$ on two different scales [268].

with the derivative of the Airy function $Ai'(X)$ and the normalized spectrometer frequency

$$\tilde{\omega} = \frac{\hbar\omega - E_g}{E_b} \left(\frac{E_b}{e\, r_B\, |\tilde{E}_0|} \right)^{2/3}. \tag{7.24}$$

Here, E_b is the exciton binding (or Rydberg) energy and r_B the exciton Bohr radius. We have taken advantage of the fact that, in a semiconductor with inversion symmetry, the sign of $\tilde{E}_0$ does not matter. The absorption spectrum is depicted in Fig. 7.23. It consists of an exponential tail for frequencies ω below the bandgap (the Franz–Keldysh effect) and an oscillatory behavior above the gap. The visibility of the oscillations decreases with increasing frequency. This has to be compared with the usual (zero-field) absorption spectrum

$$\alpha(\hbar\omega) \propto \Theta(\hbar\omega - E_g) \sqrt{\hbar\omega - E_g} \tag{7.25}$$

of a three-dimensional semiconductor within the effective-mass approximation (gray area in Fig. 7.23).

If one applies a "slowly varying" harmonic electric field rather than a true static field, i.e., $\tilde{E}_0 \rightarrow \tilde{E}_0 \cos(\omega_0 t + \phi)$, the absorption changes during an oscillation cycle $2\pi/\omega_0$ and the optical field at frequency ω sees a cycle-averaged absorption coefficient $\langle\alpha(\hbar\omega)\rangle$. This effectively averages over different values of the electric field and tends to smear out spectral structures.

In the strict static-field case, where $\omega_0 = 0$, the ratio of the ponderomotive energy and the carrier photon energy is infinity, $\langle E_{kin}\rangle/\hbar\omega_0 = \infty$. For small ω_0 still within the static-field approximation, this ratio can easily be very large compared to unity even for moderate values of the peak electric field $\tilde{E}_0$, i.e., the limit of extreme nonlinear "optics" is easily reached here.

Dynamic Franz–Keldysh effect

Up to what carrier frequencies ω_0 is the above static-field approximation valid? In Sect. 4.2 we have seen that Volkov states (rather than the Airy wave functions multiplied by $\exp(-i\hbar^{-1}E_e t)$) are solutions of the time-dependent Schrödinger equation within the effective-mass approximation for a harmonically varying electric field [213]. The energy spectrum of these Volkov states consists of a set of equidistant parabolas of energy versus electron momentum, separated by the carrier photon energy $\hbar\omega_0$ (see Fig. 4.2). Thus, for sufficiently large ω_0, these sidebands should also show up in the absorption spectrum by a series of equidistant absorption onsets [86, 87]. In practice, this will only be the case if (at least) the following four conditions are fulfilled simultaneously. (i) $\hbar\omega_0$ has to be much larger than some characteristic broadening $\hbar/T_2$ in the optical spectrum. For room-temperature conditions, $\hbar/T_2$ is typically on the order of $10\,\mathrm{meV}$ or larger. This favors large ω_0. (ii) The sidebands only acquire appreciable strength if the ratio $\langle E_{\mathrm{kin}}\rangle/(\hbar\omega_0)$ is of order unity or larger. This strongly favors small ω_0. (iii) Interband transitions must not occur, which translates into the conditions $\hbar\omega_0/E_{\mathrm{g}} \ll 1$ for the carrier frequency (again favoring small ω_0) and $\hbar\Omega_{\mathrm{R}}/E_{\mathrm{g}} \lll 1$ for the intensity. Otherwise, multiphoton interband transitions can occur (see Sects. 3.5 and 3.6.1). Using (7.3), this is equivalent to stating that the Keldysh parameter γ_{K} obeys the relation $\gamma_{\mathrm{K}} \ggg \hbar\omega_0/E_{\mathrm{g}}$, which is fulfilled for $\gamma_{\mathrm{K}} \gg 1$. (iv) At the same time, the ponderomotive energy $\langle E_{\mathrm{kin}}\rangle$ must not significantly exceed values of order $0.1\,\mathrm{eV}$, otherwise the effective-mass approximation for the electrons is no longer valid. As the hole effective masses are typically much larger than that of the electrons, the hole ponderomotive energy can be neglected to a first approximation and a single valence-band parabola results. If both $\hbar\omega_0$ and $\langle E_{\mathrm{kin}}\rangle$ are on the order of a few tenths of an electron Volt (midinfrared), the conditions (i)–(iv) are fulfilled for typical semiconductors. This also largely avoids direct excitation of longitudinal optical phonons, which would decay into acoustic phonons and eventually heat the sample, also leading to induced absorption below the bandgap. For example, the longitudinal optical phonon energy in GaAs is $36\,\mathrm{meV}$.

Samples

Experiments have been reported for $l = 350\,\mu\mathrm{m}$ thick crystalline, semi-insulating bulk GaAs ($E_{\mathrm{g}} = 1.42\,\mathrm{eV}$), for $l = 3\,\mathrm{mm}$ polycrystalline ZnSe ($E_{\mathrm{g}} = 2.7\,\mathrm{eV}$) and crystalline ZnTe ($E_{\mathrm{g}} = 2.3\,\mathrm{eV}$), all at room temperature [88]. Such rather thick samples lead to large reduced transmission below the bandgap. Above the gap, however, these samples are completely opaque and the resulting changes in absorption cannot be measured. Other authors [214] have employed a much thinner, $l = 2\,\mu\mathrm{m}$, crystalline bulk GaAs sample.

Laser systems

In order to excite the semiconductor in the midinfrared and simultaneously probe the resulting changes in sample transmission at optical frequencies near the semiconductor bandgap, 100-fs long laser pulses from a regeneratively amplified Ti:sapphire laser oscillator can be used to generate a white-light continuum in a sapphire plate, serving as the optical probe. In addition, another part of these pulses can be sent into an optical parametric amplifier. Subsequent difference-frequency mixing with the original pulses delivers tunable and intense midinfrared pulses for excitation with a duration on the order of 1 ps. At, e.g., 6 µm wavelength ($\Leftrightarrow \hbar\omega_0 = 0.2\,eV$), this duration corresponds to 50 cycles of "light".

Experiment

The anticipated changes in transmission due to the dynamic Franz–Keldysh effect should only be present if pump and probe temporally overlap in the sample. Heating effects, on the other hand, would remain if the optical probe comes after the midinfrared pump by some time delay τ. This important check has been performed [88], and heating effects could be ruled out. Figure 7.24 shows measured [214] transmission spectra for the $l = 350\,\mu m$ thick GaAs sample. The linear transmission spectrum (dashed curve) exhibits $> 40\%$ transmission below the bandgap, which is expected from accounting for the reflections on the two (uncoated) GaAs/air interfaces (each about 30%). With excitation (full curve), significant induced absorption is found for photon energies well below the GaAs absorption edge. Such an absorption into the $N = -1$ parabola of the Volkov states (see Fig. 4.2) is simply two-photon absorption, where an optical photon energy $\hbar\omega$ plus a midinfrared photon energy $\hbar\omega_0$ is larger than E_g. Under the conditions studied, this two-photon absorption is not proportional to the midinfrared intensity, as would be expected within the perturbative regime, but is rather governed by the amplitudes a_N of the Volkov sidebands in the regime $\langle E_{kin} \rangle \approx \hbar\omega_0$. However, no clear direct indication of the anticipated sidebands from the Volkov states is visible. This is in agreement with theory based on the Volkov states [88]. Keep in mind that sharp sidebands would only be expected for continuous-wave excitation or for box-shaped pulses. The sidebands are smeared out for Gaussian pulses analogous to what we have said in Sect. 3.6. For bulk semi-

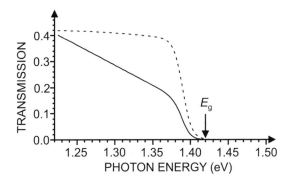

Fig. 7.24. Measured transmission spectra of bulk, $l = 350\,\mu m$ thick GaAs with (full) and without (dashed) midinfrared excitation at $\hbar\omega_0 = 0.2\,eV$ (equivalent to 6.2 µm wavelength) for an intensity of $I = 3 \times 10^9\,W/cm^2$ corresponding to $\langle E_{kin} \rangle \approx \hbar\omega_0$. [269], schematically after Ref. [88].

conductors, the onsets of N-photon absorption associated to the Volkov sidebands are generally not very pronounced anyway. Comparison with Fig. 7.23(b) shows that even the static-field approximation is able to qualitatively describe these experimental findings. Further experiments have been performed on thinner GaAs samples [214]. Here, the differential transmission $\Delta T / T$ is measured that is defined by the difference of the optical transmission with the midinfrared pump on minus that for pump off, divided by the transmission for pump off. Additional increased transparency above the bandgap has been observed [214] (not shown) – an aspect that cannot easily be understood within a perturbative approach. In the framework of the Volkov states, it is due to the upshift of the fundamental $N = 0$ parabola by the ponderomotive energy $\langle E_{kin} \rangle$ (see Fig. 4.2). Such induced transparency (reduced absorption) above the original bandgap does, however, also occur in the static-field approximation (see Fig. 7.23(b)).

For midinfrared excitation but for yet much larger intensities on the order of 10^{11} W/cm^2, mixing of optical and midinfrared photons as well as high (up to $N = 7$) harmonics of $\hbar\omega_0$ have been observed [215]. Under these conditions, the ponderomotive energy is so large (several electron Volts) that the effective-mass approximation no longer applies. Furthermore, at these intensities, the condition $\hbar\Omega_R / E_g \ll 1$ is not really fulfilled any longer, equivalent to saying that interband multiphoton transitions come into play (see our discussion on two-level systems in Sects. 3.5 and 3.6.1).

If the excitation carrier photon energy $\hbar\omega_0$ comes close to the 1s $\rightarrow$ 2p exciton transition, excitonic effects can become prominent. In GaAs quantum wells (where $E_b \approx 10$ meV), this has indeed been shown at THz frequencies ($\hbar\omega_0 = 0.5$–20 meV) [216] and at around unity exciton Keldysh parameter γ_K^x.

7.4 Photon Drag or Dynamic Hall Effect

In Sect. 4.4.1 we have seen that the photon-drag effect simply arises from the magnetic component of the Lorentz force in Newton's second law and that it can be viewed as a precursor of relativistic extreme nonlinear optics of free electrons. Loosely speaking, the photons push the electrons along the wavevector of light (for a discussion of "radiation pressure" see Refs. [217–219]). However, one should be aware that for constant intensity of light no electron acceleration results, but rather a constant drift velocity. Alternatively, the photon drag can be interpreted as the dynamic version of the Hall effect [220], where a static electric field together with a perpendicular static magnetic field lead to a motion of charges perpendicular to both of them. A nonrelativistic quantum-mechanical treatment of the photon drag gives the same result [221]. The photon-drag current must not be confused with the longitudinal component of the purely displacive photogalvanic current – both of which are proportional to the intensity of light. The latter results from the motion of charges due to optical rectification (see Sect. 2.4). In an inversion symmetric medium, the photogalvanic current is strictly zero, while the photon-drag current is generally nonzero (see our discussion in Sect. 2.5 on $\chi^{(2)}$ and $\chi_L^{(2)}$).

In Problem 4.4 we have further seen that the inevitable damping of crystal electrons, due to the microscopic scattering processes in solids, only leads to a modified prefactor. In the perturbative regime (where $\mathcal{E}^2 \ll 1$), the resulting photon-drag current density j_{pd} according to (4.55) is proportional to the intensity of light and scales as $\mathcal{E}^2$ (the dimensionless field strength $\mathcal{E}$ is defined in (4.44)). As $\mathcal{E}^2 \propto 1/\omega_0^2$ and $\mathcal{E}^2 \propto 1/m_e^2$, the effect is large for infrared carrier frequencies ω_0 and for semiconductors with small effective electron (rest) masses m_e (see Table 4.1). In this regime, the photon-drag effect is used in commercially available, high-speed, room-temperature, infrared germanium photodetectors, which do not require an external bias voltage. Typically, they are sold for the 10.6 µm wavelength of CO_2 lasers.

Note that one is indeed restricted to the perturbative regime $\mathcal{E}^2 \ll 1$ in semiconductors. For example, for $m_e/m_0 = 0.1$ and independent of the laser carrier frequency ω_0, $\mathcal{E}^2 = 10^{-4}$ would correspond to a ponderomotive energy of $\langle E_{kin} \rangle = 1.27\,eV$ according to (4.76) (here we have to replace m_0 by m_e). This value is already comparable to the width of the conduction band and certainly no longer compatible with the effective-mass approximation (see Sect. 4.3). Thus, long before one approaches the nonperturbative regime with $\mathcal{E}^2 \approx 1$, one rather either enters the case of carrier-wave Bloch oscillations described in Sect. 4.3.2 or one gets beyond the damage threshold of the sample.

The total photon-drag current density in a semiconductor is the sum of the electron and the hole contribution. According to (4.55), the two have opposite sign, because the current density is proportional to the cube of the charge. Furthermore, introducing a finite (normalized) damping time τ for electrons (e) and holes (h) according to Problem 4.4, we obtain

$$j_{pd} = \left(-\underbrace{\frac{\tau_e^2}{1+\tau_e^2}\frac{N_e\,e^3}{2\,V m_e^2\,c_0\,\omega_0^2}}_{\geq 0} + \underbrace{\frac{\tau_h^2}{1+\tau_h^2}\frac{N_h\,e^3}{2\,V m_h^2\,c_0\,\omega_0^2}}_{\geq 0} \right) \tilde{E}_0^2 \gtreqless 0 \qquad (7.26)$$

in analogy to (4.55). Thus, if the first term in (7.26) cancels with the second one, the total current density j_{pd} would clearly be zero – a coincidence, which does not occur in practice, because usually doped samples are employed (see Fig. 7.25) in which one component overwhelms the other. To actually measure the current, Ohmic contacts have to be fabricated on the surfaces of a crystal and short-circuit conditions have to be employed. On the other hand, under open-circuit conditions, the photon drag leads to a spatial separation of electrons and holes, hence to a build-up of an electric field, the so-called drag field.

The drag field or the drag current, respectively, are not always parallel to the wavevector of the exciting light [223]. Microscopically, this effect arises because of the anisotropy of the electron and hole effective masses. This anisotropy is exemplified for m_e of germanium in Table 4.1. Loosely speaking, the electron or hole drift direction becomes a compromise between the wavevector of light and directions of small effective mass. The general mathematical form for the photon-drag current-density vector is given by (see Problem 7.2)

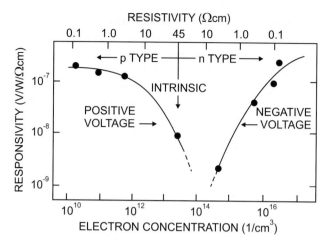

Fig. 7.25. Measured responsivity of eight different germanium photon-drag photodetectors (dots) versus electron concentration under open-circuit conditions. The LHS corresponds to p-type material, the RHS to n-type. Reprinted from Ref. [222] by permission of the American Institute of Physics.

$$\boldsymbol{j}_{pd} = \begin{pmatrix} j_1 \\ j_2 \\ j_3 \end{pmatrix} \quad \text{with} \quad j_i = \sum_{j,k,l=1}^{3} \mathcal{T}_{ijkl} \, \hat{K}_j \tilde{E}_k \tilde{E}_l \,, \tag{7.27}$$

where $\hat{\boldsymbol{K}} = \boldsymbol{K}/|\boldsymbol{K}|$ is the unit wavevector of light, $\mathcal{T}$ is a tensor of rank four and the linearly polarized laser electric-field vector is expressed as

$$\boldsymbol{E}(t) = \begin{pmatrix} \tilde{E}_1 \\ \tilde{E}_2 \\ \tilde{E}_3 \end{pmatrix} \cos(\omega_0 t + \phi) \,. \tag{7.28}$$

For example, in a cubic medium with inversion symmetry (such as germanium), the only nonvanishing elements of the photon-drag tensor are [224]

$$\mathcal{A} = \mathcal{T}_{1111} = \mathcal{T}_{2222} = \mathcal{T}_{3333} \,,$$
$$\mathcal{B} = \mathcal{T}_{1122} = \mathcal{T}_{1133} = \mathcal{T}_{2233} = \mathcal{T}_{3311} = \mathcal{T}_{3322} = \mathcal{T}_{2211} \,,$$
$$\mathcal{C} = \mathcal{T}_{2323} = \mathcal{T}_{3131} = \mathcal{T}_{1212} \,, \tag{7.29}$$

and $\mathcal{T}_{ijkl} = \mathcal{T}_{jikl} = \mathcal{T}_{ijlk} = \mathcal{T}_{jilk}$. Taking advantage of the fact that the electric-field vector and the wavevector of light are perpendicular to one another, i.e., $\tilde{\boldsymbol{E}} \cdot \hat{\boldsymbol{K}} = 0$, (7.27) simplifies to

$$j_i = \hat{K}_i \left(\mathcal{B} \left(\tilde{E}^2 - \tilde{E}_i^2 \right) + (\mathcal{A} - 2\mathcal{C}) \tilde{E}_i^2 \right) \,, \tag{7.30}$$

with $\tilde{E}^2 = \tilde{E}_1^2 + \tilde{E}_2^2 + \tilde{E}_3^2$. We are left with only the two independent material parameters $\mathcal{B}$ and $(\mathcal{A} - 2\mathcal{C})$.

Let us consider two examples of geometries. (i) For $\hat{\boldsymbol{K}} = (0, \cos\varphi, \sin\varphi)^{\mathrm{T}}$ and our "usual" choice of $\tilde{\boldsymbol{E}} = \tilde{E} (1, 0, 0)^{\mathrm{T}}$, we immediately obtain $\boldsymbol{j}_{\mathrm{pd}} \| \boldsymbol{K}$ for any value of φ, $\mathcal{B}$, and $(\mathcal{A} - 2\mathcal{C})$. (ii) For the geometry $\hat{\boldsymbol{K}} = (0, \cos\varphi, \sin\varphi)^{\mathrm{T}}$ and $\tilde{\boldsymbol{E}} = \tilde{E} (0, \sin\varphi, -\cos\varphi)^{\mathrm{T}}$ we get

$$\begin{pmatrix} j_1 \\ j_2 \\ j_3 \end{pmatrix} = \tilde{E}^2 \mathcal{B} \begin{pmatrix} 0 \\ \cos\varphi \, (\cos^2\varphi + \mathcal{D} \sin^2\varphi) \\ \sin\varphi \, (\sin^2\varphi + \mathcal{D} \cos^2\varphi) \end{pmatrix}, \qquad (7.31)$$

where we have introduced the dimensionless parameter $\mathcal{D} = (\mathcal{A} - 2\mathcal{C})/\mathcal{B}$. For φ equal to 0, $\pi/4$, $\pi/2$, ... , we again obtain $\boldsymbol{j}_{\mathrm{pd}} \| \boldsymbol{K}$ for any value of $\mathcal{D}$. Generally, the photon-drag current and the wavevector of light include an angle θ, which depends only on $\mathcal{D}$ for a given value of φ. The angle θ can be determined via the dot product $\boldsymbol{j}_{\mathrm{pd}} \cdot \boldsymbol{K} = |\boldsymbol{j}_{\mathrm{pd}}| |\boldsymbol{K}| \cos\theta$. For $\mathcal{D} = 1$, we obtain $\theta = 0$ ($\Leftrightarrow \boldsymbol{j}_{\mathrm{pd}} \| \boldsymbol{K}$) for all φ. For $\mathcal{D} \to 0$, the maximum angle is $\theta = 0.340$ (19.5 degrees) for, e.g., $\varphi = 0.478$ (27.4 degrees). For $\mathcal{D} \to \infty$, $\theta = \pi/2$ is possible mathematically, i.e., the photon-drag current can *in principle* even be perpendicular to the wavevector of light. For p-type germanium, actual angles as large as $\theta = 0.157$ (9.0 degrees) have been determined on the basis of experimental data [224].

We have already seen in Sects. 2.5 and 4.4 that the photon drag is closely connected with a "longitudinal" component of the optical polarization oscillating at carrier frequency $2\omega_0$. In a crystal with reduced symmetry, in analogy to the photon drag in (7.27), this contribution can be generalized according to

$$\boldsymbol{P}(t) = \begin{pmatrix} \tilde{P}_1 \\ \tilde{P}_2 \\ \tilde{P}_3 \end{pmatrix} \sin(2\omega_0 t + 2\phi) \quad \text{with} \quad \tilde{P}_i = \sum_{j,k,l=1}^{3} \tilde{T}_{ijkl} \, \hat{K}_j \tilde{E}_k \tilde{E}_l . \qquad (7.32)$$

Here we have assumed negligible damping. With damping, an additional phase shift comes into play. This form of the optical polarization can even exhibit a transverse component of second-harmonic generation in an inversion-symmetric crystal, because $\boldsymbol{P}$ and $\boldsymbol{K}$ may include an angle θ.

Problem 7.1. Discuss (7.27) under space inversion.

Problem 7.2. Derive (7.27) from Newton's second law in analogy to our discussion in Sect. 4.4.1 but using an electron (hole) effective-mass tensor rather than an isotropic effective mass.

7.5 Conical Second-Harmonic Generation

The "longitudinal" contribution to second-harmonic generation (SHG) associated with the photon drag allows for SHG in an isotropic solid. A distinct mechanism that does the same thing is conical harmonic generation [225]. In contrast to the photon drag, it is mediated by a $\chi^{(5)}$ process to lowest order. Here, the difference-frequency mixing of four photons from the fundamental wave with carrier frequency ω_0 and one photon from the SHG field with carrier frequency $2\omega_0$ leads to a contribution of the optical polarization at frequency $4\omega_0 - 2\omega_0 = 2\omega_0$. This contribution is proportional to the fourth power of the fundamental envelope and proportional to the envelope of the SHG field itself and drives the SHG field via the wave equation. Within the slowly varying envelope approximation (see Sect. 6.2), we can closely follow along the lines of (6.7), (6.8) and (6.9), which assume negligible depletion for the fundamental wave ($N = 1$). We obtain for the SHG with $N = 2$

$$2 i K_2 e^{iK_2 z} \frac{\partial \tilde{E}_2}{\partial z}(z, t) = -\mu_0 (2\omega_0)^2 \epsilon_0 \chi_{SHG}^{(5)} \tilde{E}_1^4 e^{i4K_1 z} \tilde{E}_2^* e^{-iK_2 z} . \qquad (7.33)$$

Here we have introduced $\chi_{SHG}^{(5)} = \chi^{(5)} 5/2^5$ with $\chi^{(5)}$ defined via (2.37). In the case of phase matching, i.e., $\Delta K = 4K_1 - 2K_2 = 0$ equivalent to $c(\omega_0) = c(2\omega_0)$, this leads to

$$\frac{\partial \tilde{E}_2}{\partial z} = +i g \tilde{E}_2^* , \qquad (7.34)$$

where we have lumped the various prefactors into the real coefficient $g \propto \chi_{SHG}^{(5)} \tilde{E}_1^4$. It is straightforward to find the general solution

$$\tilde{E}_2(z, t) = \mathcal{A}_+ (1 + i) e^{+gz} + \mathcal{A}_- (1 - i) e^{-gz} . \qquad (7.35)$$

The real coefficients $\mathcal{A}_+$ and $\mathcal{A}_-$ can be determined from the initial condition $\tilde{E}_2(0, t) = (\mathcal{A}_+ + \mathcal{A}_-) + i(\mathcal{A}_+ - \mathcal{A}_-)$. For the special case $\mathcal{A}_+ = \mathcal{A}_-$, this simplifies to

$$\tilde{E}_2(z, t) = \tilde{E}_2(0, t) \left(\cosh(gz) + i \sinh(gz)\right) . \qquad (7.36)$$

In any case, for large propagation coordinates z with $|gz| \gg 1$, we get an exponential growth of the SHG field envelope $\tilde{E}_2$, as well as of the SHG intensity $I_{SHG}(l) \propto |\tilde{E}_2(z = l, t)|^2$ with medium thickness l. For the special case $\mathcal{A}_+ = \mathcal{A}_-$ we obtain

$$I_{SHG}(l) = I_{SHG}(0) \left(\cosh^2(gl) + \sinh^2(gl)\right) . \qquad (7.37)$$

This is in sharp contrast to the usual quadratic growth $I_{SHG} \propto l^2$ of phase-matched SHG from a $\chi^{(2)}$ process in the low pump depletion approximation (see (6.10)). For $\tilde{E}_2(0, t) = 0$, i.e., without any seed of the SHG at the front of the sample, the SHG would still be strictly zero. Any small seed, on the other hand, enables appreciable SHG signals for large gain, i.e., for $|gl| \gg 1$. In a quantum optical treatment, quantum fluctuations may provide such seeding.

Note, however, that the phase-matching condition $c(\omega_0) = c(2\omega_0)$ is generally *not* fulfilled in dispersive media. Usually, one rather has $c(\omega_0) > c(2\omega_0)$. In traditional nonlinear optics, birefringence comes to the rescue, which is obviously absent for isotropic media. Still, the condition $4K_1 - 2K_2 = 0$ can be fulfilled if the wavevectors $\boldsymbol{K}_1 || (0, 0, 1)^T$ with $|\boldsymbol{K}_1| = \omega_0/c(\omega_0)$ and $\boldsymbol{K}_2$ with $|\boldsymbol{K}_2| = 2\omega_0/c(2\omega_0)$ are *not* parallel, but rather include an angle θ. If their z-components obey the condition $4(K_1)_z - 2(K_2)_z = 4|K_1| - 2|K_2|\cos\theta = 0$, this immediately translates into

$$\cos\theta = \frac{c(2\omega_0)}{c(\omega_0)}, \qquad (7.38)$$

which can be fulfilled if the RHS ≤ 1 but not if the RHS > 1. This means that the SHG lies on a cone with opening angle θ along the wavevector of the fundamental, hence the name "conical SHG". At the exit surface of the material, one expects a bright ring. This is illustrated in Fig. 7.26.

In general, such conical phase matching allows for any odd or even harmonic of order N from a $\chi^{(2N+1)}$ susceptibility via the difference-frequency mixing of the sum of $2N$ fundamental photons with one N-th harmonic photon.

Conical SHG has not been observed, while conical third-harmonic generation via a $\chi^{(7)}$ process has indeed been observed [225]. Here, $\chi^{(3)}$-based third-harmonic generation (not phase matched) delivers the required seed. In these experiments, sapphire is excited by 10-μJ energy, 50-fs duration pulses centered at 1.5 μm wavelength, leading to an opening angle θ of 12 degrees. Estimating a spot radius of 10 μm on the basis of Ref. [225], these numbers correspond to a peak intensity on the order of $I = 6 \times 10^{13}$ W/cm^2, equivalent to $\tilde{E}_0 = 2 \times 10^{10}$ V/m in vacuum.

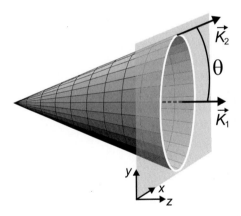

Fig. 7.26. Illustration of the emission characteristics of conical second-harmonic generation (SHG) in an isotropic material. The SHG wavevector K_2 includes an angle θ with the fundamental wavevector K_1 (central axis). The depicted angle of $\theta = 0.35$ (20 degrees) corresponds to $c(2\omega_0) = 0.94\,c(\omega_0)$. To lowest order, conical SHG is a fifth-order process.

8

Extreme Nonlinear Optics of Atoms and Electrons

The theoretical basis for the extreme nonlinear optics of atoms, free electrons, and the vacuum has already been laid in Sects. 5.4, 4.4 and 4.5, respectively. We have seen that, for atoms, the regime of extreme nonlinear optics is entered if the ponderomotive energy becomes comparable to the binding energy ($\Leftrightarrow \gamma_K \approx 1$). For free electrons, the ponderomotive energy needs to approach the electron rest energy ($\Leftrightarrow |\mathcal{E}| \approx 1$). To generate real electron–positron pairs from vacuum, the electron rest energy has to be matched by the potential drop over the Compton wavelength. Corresponding typical laser intensities have turned out to be 10^{14}, 10^{18}, and 10^{30} W/cm^2, respectively.

Laser systems

To reach such intensities, amplified laser systems are necessary. In particular, the concept of *chirped-pulse amplification* has led to enormous progress [16, 17]. Without this "trick", the laser intensity (not the pulse energy) soon reaches a point at which it destroys the optical components on the way. In chirped-pulse amplification, the pulses are thus first stretched in time by a few orders of magnitude (ideally, this leaves their amplitude spectrum unaffected), then amplified to large pulse energies while keeping the peak intensity below the damage threshold of the gain medium, and finally recompressed in time.

8.1 High-Harmonic Generation From Atoms

In Sect. 5.4 we have seen that the interaction of atoms with laser pulses can lead to the generation of high harmonics up to harmonic orders N of several hundreds. These high harmonics can result in attosecond pulses [226, 227] or attosecond pulse trains [228, 229] of extreme ultraviolet (EUV) radiation. Figure 8.1 shows three different geometries that are used in experiments. For a direct comparison with microscopic calculations, e.g., on the basis of the time-dependent Schrödinger equation, one would prefer isolated dipoles and try to avoid propagation effects as much as possible. For a dilute gas, geometry (a) in Fig. 8.1 can come near to this goal. In contrast to this, if one

(a)

(b)

(c)

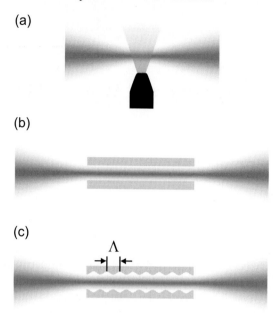

Fig. 8.1. Geometries for high-harmonic generation from atoms (schematically, not to scale). **(a)** The laser pulse interacts with the atoms of a gas jet (a geometry used by many authors), **(b)** the atoms in a hollow gas capillary [230, 231] enabling phase matching or **(c)** with the atoms in a modulated capillary [232] allowing for quasi phase-matching. Λ is the modulation period [269].

is rather interested in obtaining a maximum EUV intensity, large interaction lengths are desired. We have already seen in our discussion based on the one-dimensional version of the slowly varying envelope approximation in Sect. 6.2 that the coherence length $l_{coh}(N)$ according to (6.12) sets an upper limit. In practice, additional effects such as, e.g., reabsorption, self-defocusing or the dispersion of the Gouy phase further reduce the effective coherence length. It has been argued on the basis of detailed calculations [135] that geometry (a) is still advantageous for few-cycle pulses, whereas for longer pulses geometries (b) or (c) are more appropriate. Still, present typical best high-harmonic conversion efficiencies are around 10^{-5}. This has to be compared with the near-unity conversion efficiency achieved for second-harmonic generation in traditional nonlinear optics.

Let us have a closer look at geometry (a) in Sect. 8.1.1, at (b) in Sect. 8.1.2 and at (c) in Sect. 8.1.3

8.1.1 Gas Jets

Figure 8.2 gives one example of a high-harmonic spectrum from helium excited by 200-fs pulses, revealing many well-resolved harmonic orders. The cutoff harmonic order (here $N_{cutoff} \approx 95$) discussed in Sect. 5.4.1 is clearly visible. More experimental

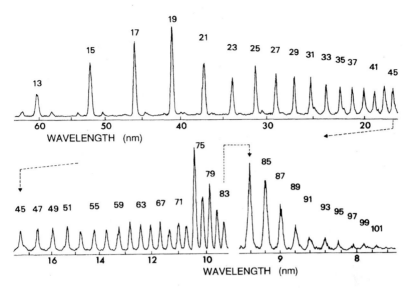

Fig. 8.2. Example of high-harmonic generation. Here, helium at 20 000 Pa (200 mbar) pressure is excited by 200-fs pulses at $\hbar\omega_0 = 1.6\,eV$ and $I = 3 \times 10^{15}$ W/cm^2. The geometry corresponds to Fig. 8.1(a). The harmonics are labeled by their order N. The small unmarked peaks result from second-order diffraction of the grating of the spectrometer. Reprinted figure with permission from K. Miyazaki and H. Takada, Phys. Rev. A **52**, 3007 (1995) [242]. Copyright (1995) by the American Physical Society.

examples can be found in Refs. [233–244] and in the reviews [245, 246]. Using helium, large photon energies [243, 244] have been generated with $\hbar\omega \approx 0.5\,keV$ (2.5 nm wavelength) at $\hbar\omega_0 = 1.5\,eV$ laser carrier photon energy, corresponding to a harmonic order of $N \approx 333$. This is consistent with the anticipated linear dependence of the cutoff harmonic order N_{cutoff} versus ionization potential E_b according to (5.19) and the fact that helium has the largest ionization potential of the gases shown in Table 5.1. The occurrence of harmonic orders much larger than those in Fig. 8.2 at *lower* pulse intensities is enabled by the fact that the authors [243, 244] have used much shorter pulses. According to our discussion in Sect. 5.4.1, this allows for large electron ponderomotive energies $\langle E_{kin} \rangle$ in the first optical cycle after ionization at low ionization degrees (compare with Fig. 5.12 and (5.19)). More recently, even photon energies larger than 0.7 keV have been obtained from helium [247] by (partial) phase matching via the frequency-dependence of the Gouy phase (see Chap. 6.3).

To understand the shape of the high-harmonic spectra in detail and to optimize conversion efficiencies, one has to consider phase-matching effects. We start from our discussion in Sect. 6.2, which is based on a one-dimensional version of the slowly varying envelope approximation. We have seen that the coherence length of the N-th harmonic (see (6.12))

$$l_{coh}(N) = \frac{\pi}{|\Delta K|} \tag{8.1}$$

is related to the mismatch ΔK of the fundamental wavevector of light and that of the N-th harmonic according to (6.11). This expression can easily be rewritten in terms of the refractive index $n(\omega)$ as

$$\Delta K = \frac{N\omega_0}{c_0} \left(n(\omega_0) - n(N\omega_0) \right) . \tag{8.2}$$

The total mismatch ΔK comprises a contribution ΔK_{atom} of atoms that are not ionized, a free-electron part ΔK_e for the electrons of ionized atoms and – in the case that a waveguide is used – the properties of the empty waveguide ΔK_{wg}. Mathematically, strictly speaking, one needs to add up the three corresponding linear optical susceptibilities χ_i and then compute the total refractive index difference in (8.2) from that. This procedure can be simplified by the fact that all relevant individual refractive indices are very close to unity, in which case we can approximate $n(\omega) = \sqrt{\epsilon(\omega)} = \sqrt{1 + \sum_i \chi_i(\omega)} \approx 1 + \sum_i \chi_i(\omega)/2 \approx 1 + \sum_i (n_i(\omega) - 1)$. Correspondingly, the wavevector mismatch can be written as the sum

$$\Delta K = \Delta K_{atom} + \Delta K_e + \Delta K_{wg} . \tag{8.3}$$

Here, we have tacitly assumed that the refractive index of the atoms and/or of the free electrons does not influence the waveguide properties, consistent with the above approximation. The refractive index of the atoms that are not ionized is given by $n_{atom} = \sqrt{\epsilon} = \sqrt{1 + \chi}$. This susceptibility χ generally consists of a series of resonances. Equation (3.3) (also see Fig. 3.1) considers just one at transition frequency Ω. Typically ($\omega_0 < \Omega < N\omega_0$), the index is larger than unity for the fundamental and very close to unity for the high harmonics. Thus, we usually have

$$\Delta K_{atom} > 0 . \tag{8.4}$$

The magnitude of this contribution obviously increases with increasing gas pressure ($\propto N_{atom} = N_{osc}$ in (3.3)). For a given gas pressure, it decreases with increasing degree of ionization of the atoms, i.e., with increasing light intensity. At the same time, the free-electron contribution increases in magnitude. Its refractive index $n_e = \sqrt{\epsilon} = \sqrt{1 + \chi}$ can be computed via the free-electron or Drude susceptibility χ according to (4.2) (also see Fig. 4.1). Typically, the electron density is below 10^{20} /cm^3 and the plasma frequency ω_{pl} from (4.3) is still much smaller than the fundamental carrier frequency ω_0, hence $0 < \epsilon < 1$ and the refractive index of the fundamental is real, positive and smaller than unity. The index is again very nearly unity for the high harmonics. Consequently, we have

$$\Delta K_e < 0 . \tag{8.5}$$

If one is interested in generating very high harmonics, the cutoff order N_{cutoff} has to be large, thus large ponderomotive energies $\langle E_{kin} \rangle \propto I$ or intensities I are required according to (5.19), hence many atoms are ionized and this free-electron contribution can become quite prominent, especially for pulses containing several or many cycles of light (see Fig. 5.12). *Note that the free-electron density increases with time during the pulse, which means that the phase-matching condition changes with time as well.*

▶ **Example 8.1.** Suppose that the free-electron contribution in (8.3) dominates. We consider an electron density $N_e/V = 10^{18}$ /cm^3, which corresponds to a plasma energy (4.3) of $\hbar\omega_{pl} = 0.037\,eV$. For a fully singly ionized gas, the corresponding gas pressure would be 4141 Pa (41 mbar) at $T = 300$ K according to the ideal gas equation (see corresponding discussion in Sect. 4.4.2). The harmonic order of interest be $N = 101$ and $\hbar\omega_0 = 1.5\,eV$. With (4.2) this leads to a fundamental refractive index of $n_e(\omega_0) = 0.9997$ and a high-harmonic index of $n_e(N\omega_0) = 1.00000$. Thus, we obtain a coherence length of $l_{coh}(101) = 13.6\,\mu$m. For $N \gg 1$, this value scales $\propto 1/N$. ◀

Two additional effects that are beyond our one-dimensional plane-wave approximation also influence the high-harmonic spectra. Above, we have already mentioned the dispersion of the Gouy phase, which has been discussed in Sect. 6.3. The underlying physics is related to that of the gas capillaries, which we will discuss below. Furthermore, the transverse (e.g., Gaussian) distribution of the beam in the focus leads to large electron densities in the center – where the intensity and the degree of ionization are high – and low densities in the wings. Thus, the free-electron refractive index is lower in the center than in the wings, which leads to defocusing of the beam. This mechanism effectively reduces the interaction length.

8.1.2 Hollow Waveguides

Let us distinguish between two cases, (i) and (ii). For (i), the free-electron contribution in (8.3) will be negligible with respect to the atomic contribution, i.e. $|\Delta K_{atom}| \gg |\Delta K_e|$. This is realistic if one only wants to generate rather "low" harmonic orders. In contrast to this, the free-electron contribution will be much larger than the atomic one for (ii), i.e. $|\Delta K_{atom}| \ll |\Delta K_e|$, corresponding to high intensities and high harmonic orders N. (i) is discussed in this section, (ii) in the next.

In case (i), the waveguide dispersion has to compensate for the positive atomic contribution, i.e., $\Delta K_{atom} > 0$ has to be cancelled by some $\Delta K_{wg} < 0$ to give $\Delta K = 0$ or at least $\Delta K \approx 0$. This can be accomplished by using a hollow gas capillary (see Fig. 8.1(b)). The light propagating in such a capillary is not really confined to the waveguide – the waveguiding is lossy. Nevertheless, in the language of geometrical optics, rays impinging onto the glass capillary (typical refractive index equals 1.5) under nearly grazing incidence can be reflected with just a few per cent loss per reflection. To compute the field distribution and, more importantly, to discuss the dispersion relation of light propagating along such a waveguide, one needs to employ a wave picture. This electromagnetic problem was solved many years ago [248]. The resulting solutions can be expressed in terms of Bessel functions.

Intuitively, we can closely follow along the lines of our reasoning for the Gouy phase in Sect. 6.3: The transverse confinement of the waveguide mode leads to a spread in transverse momentum via the "uncertainty" relation. This, together with the fact that the modulus $|K| = \sqrt{K_x^2 + K_y^2 + K_z^2} = \omega/c_0$ is constant, reduces K_z, the component along the capillary axis. Appropriately averaging over the transverse wavefront delivers an effective K_z^{eff}, the propagation constant. The effective refractive index n_{wg} of the empty waveguide results from the dispersion relation of light

$$\frac{\omega}{K_z^{\mathrm{eff}}} = c = \frac{c_0}{n_{\mathrm{wg}}(\omega)} . \tag{8.6}$$

As $K_z^{\mathrm{eff}} < |K|$, we can conclude that $n_{\mathrm{wg}}(\omega) < 1\ \forall \omega$ – the phase velocity of light in the waveguide is larger than the vacuum speed of light c_0. Indeed, the complete calculation [248] delivers a frequency dependence according to

$$n_{\mathrm{wg}}(\omega) = 1 - \frac{\omega_{\mathrm{crit}}^2}{\omega^2} . \tag{8.7}$$

For large frequencies ω, corresponding to wavelengths much smaller than the diameter of the capillary, the light field hardly "feels" the waveguide (this is the limit of geometrical optics), hence we have $n_{\mathrm{wg}}(\omega \to \infty) \to 1$. Inserting the index according to (8.7) into (8.2), we obtain

$$\Delta K_{\mathrm{wg}} < 0 . \tag{8.8}$$

The modulus of ΔK_{wg} can be tailored by the radius of the capillary, r_{cap}, because $\omega_0 \gg \omega_{\mathrm{crit}} \propto 1/r_{\mathrm{cap}}$. Remember that the atomic contribution $\Delta K_{\mathrm{atom}} > 0$ depends on the gas pressure. Thus, the phase-matching condition $\Delta K = \Delta K_{\mathrm{atom}} + \Delta K_{\mathrm{wg}} = 0$ can be achieved by *pressure tuning*. This is schematically illustrated in Fig. 8.3.

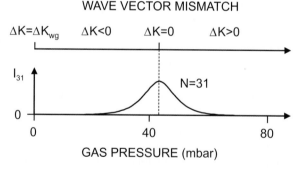

Fig. 8.3. Intensity (linear scale) of the $N = 31$st harmonic versus argon gas pressure in a hollow waveguide (see Fig. 8.1(b)). The pressure tuning curve is drawn schematically after Ref. [230]. They used 20-fs pulses at $\hbar\omega_0 = 1.55\,eV$ and $I = 2 \times 10^{14}\,W/cm^2$ to excite argon within a 150-μm diameter capillary a few centimeters in length l. Naively, one might expect to observe the side lobes of the $\propto \mathrm{sinc}^2\left(\frac{\Delta K l}{2}\right)$ phase-matching function according to (6.10), where ΔK according to (8.3) is a linear function of the pressure. However, the combined effect of absorption and varying levels of ionization smears out these structures and broadens the peak. Hence, the ΔK quoted on the top have to be interpreted as effective average quantities.

It has been shown that by using such hollow waveguides, the spatial coherence of the high harmonics also improves [249].

8.1.3 Quasi Phase-Matching in Modulated Capillaries

For the conditions of very high harmonic generation, the free-electron contribution ΔK_e becomes very prominent and overwhelms ΔK_{atom} (case (ii) from Sect. 8.1.2). $\Delta K_e < 0$ can clearly *not* be cancelled by a hollow waveguide with $\Delta K_{wg} < 0$. This would only make matters worse. Only two years after the birth of the Ruby laser, i.e., in 1962, Bloembergen and coworkers [250] introduced the concept of quasi phase-matching, which comes to the rescue. Suppose that we can not get rid of a finite wavevector mismatch $\Delta K < 0$. If the fundamental and the harmonic wave have propagated over length l, the acquired phase difference equals $\Delta K\, l$. For the coherence length $l = l_{coh}$, this phase difference is π. If one lets the waves propagate over the next length $l = l_{coh}$ under the same conditions, the harmonic intensity would be strictly zero thereafter (see (6.10)). If, however, one could reverse the sign of the nonlinearity for the next length $l = l_{coh}$, i.e., replace $\tilde{\chi}^{(N)} \rightarrow -\tilde{\chi}^{(N)}$, constructive interference is recovered. The resulting spatial period

$$\Lambda = 2\,l_{coh} \qquad (8.9)$$

can then be repeated infinitely. This was the original idea of quasi phase-matching. It does not work here. A related idea that does work, although less efficient, is to periodically modulate the magnitude of high-harmonic generation [232]. In Sects. 5.1 and 5.3 we have seen that high-harmonic generation is extremely sensitive with respect to the laser intensity and/or the CEO phase ϕ. Thus, even a minor modulation in the radius of the gas capillary can essentially switch on and off the high-harmonic generation (see Fig. 8.1(c)). This behavior is inherently nonperturbative (see Sect. 5.3) and cannot really be described by nonlinear optical susceptibilities. Still, if the fundamental envelope $\tilde{E}_1$ in (6.8) is reduced by merely 2%, the term $\tilde{\chi}^{(N)}\, \tilde{E}_1^N$ changes by a factor of 29 for the harmonic order $N = 167$. Suppose that $\tilde{E}_1 \approx$ const. for propagation over length $l = l_{coh}$, hence high harmonics are generated. For the next length $l = l_{coh}$, $\tilde{E}_1^N \approx 0$, thus the high-harmonic intensity does not grow in the second half of the period but the phase difference after one period $\Lambda = 2\,l_{coh}$ equals $\pi + \pi = 2\pi$ and constructive interference is recovered. From integration of (6.8) with $\tilde{E}_1^N = \tilde{E}_1^N(z)$ it can easily be seen that this idea also works for an arbitrary periodic dependence $\tilde{E}_1(z)$ as long as the period is given by $\Lambda = 2\,l_{coh}$. For a fixed modulation period Λ, the condition $\Lambda = 2\,l_{coh}$ can be course-adjusted by the gas pressure via the dependence of $l_{coh} = \pi/|\Delta K|$ on the density of atoms, hence of free electrons. As already pointed out above, the number of free electrons $N_e(t)$ increases monotonically with time t during the pulse (see Fig. 5.12) and the condition $\Lambda = 2\,l_{coh}(N_e(t)/V)$ is met only for a small time interval within the laser pulse. In principle, this allows isolated attosecond EUV pulses to be generated rather than pulse trains (compare Fig. 5.1) even with optical pulses containing many cycles of light.

Using such modulated waveguides, Ref. [251] has recently demonstrated rather efficient generation of photon energies up to $\hbar\omega = 250\,eV$ from *highly ionized* argon corresponding to a harmonic order of about $N = 167$ at $\hbar\omega_0 = 1.5\,eV$. This also takes advantage of the fact that the ionization potential of Ar^+ ions ($E_b = 27.6\,eV$) is larger than that of Ar ($E_b = 15.8\,eV$). According to (5.19), this leads to a larger cutoff. They [251] use incident pulses of 22 fs duration with a peak intensity of $I = 1.3 \times 10^{15}\,W/cm^2$, a gas pressure of 933 Pa (9 mbar) and a modulation period of $\Lambda = 250\,\mu m$. Typically, the diameter of the capillary is modulated by 5–10% only (see Fig. 8.1(c)).

8.1.4 Dependence on the Carrier-Envelope Phase

High-harmonic generation from few-cycle optical pulses depends on the value of the carrier-envelope offset (CEO) phase ϕ. We have already seen that this dependence can be understood on different levels of sophistication. It appears within the phenomeno-logical approach based on nonlinear optical susceptibilities via an interference of spectrally adjacent harmonics (see Fig. 5.2). Within the electrostatic tunneling approximation in Sect. 5.3, the CEO-phase dependence arises from different values of the actual instantaneous laser electric field (see Fig. 5.6). Historically, a dependence on the CEO phase has been addressed theoretically in Refs. [252, 253] and was later observed experimentally in Ref. [254].

Laser systems

In these experiments [254], the authors used a CEO-phase stabilized mode-locked laser oscillator, amplified the pulses to few mJ pulse energy, generated a "white-light" continuum via self-phase modulation in a hollow-core waveguide filled with neon, recompressed the resulting pulses and used these 5-fs pulses for excitation of a 2-mm-long sample of neon gas at intensities around $I = 7 \times 10^{14}\,W/cm^2$. Amazingly, the CEO phase of these excitation pulses turns out to be also fixed if the CEO phase of the mode-locked oscillator is fixed with a relative jitter of merely 50 mrad, equivalent to less than 1% of 2π. The actual value of ϕ, on the other hand, changes by several times 2π from the oscillator towards the gas sample.

Experiment

Measured high-harmonic spectra are depicted in Fig. 8.4. The observed behavior is periodic in ϕ with π (rather than 2π) periodicity because of the inversion symmetry of the gas. For CEO phase $\phi = 0, \pi$ (see (b)), the high-harmonic peaks on the high-energy end of the spectrum (120–130 eV) merge into a continuum as expected from Fig. 5.2, whereas they are clearly separated for CEO phases $\phi = \pm\pi/2$. In contrast to Fig. 5.2, however, the peaks at lower photon energy do not merge for any value of ϕ in Fig. 8.4. Indeed, Fig. 5.6 reminds us that the photon energies near the cutoff are special in that they are almost exclusively generated in the central cycle of the pulse. Also note that the peaks between 90 and 120 eV move with ϕ, i.e., they are not always centered at photon energies $\hbar\omega_N = N\,\hbar\omega_0$ with odd integer N. As a result,

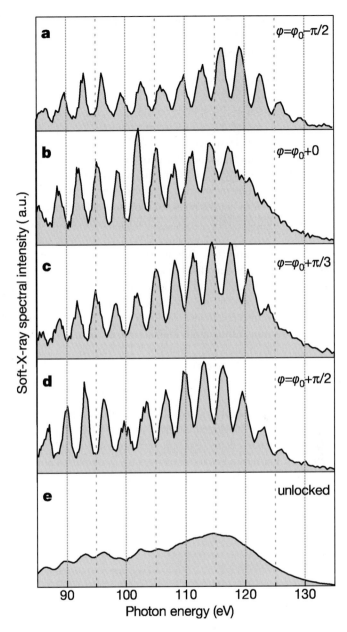

Fig. 8.4. Measured EUV spectra from neon at 16 000 Pa (160 mbar) pressure. Excitation is with 5-fs pulses around $\hbar\omega_0 = 1.5\,e$V and at an estimated intensity of $I = 7 \times 10^{14}\,\text{W/cm}^2$. **(a)**–**(d)** correspond to different but fixed CEO phases (called φ in this work) as indicated, in **(e)** the CEO phase is not stabilized. The phase offset φ_0 is not known. Compare with Figs. 5.2 and 5.6. Reprinted by permission from Nature [254] copyright (2003) Macmillan Publishers Ltd.

the peaks are smeared out if ϕ is not stabilized (see Fig. 8.4(e)). On the other hand, the peaks are equidistant with a spacing close to $2\,\hbar\omega_0 = 3\,eV$ and we can express the high-harmonic peak positions[1] with odd N as

$$\omega_N = N\,\omega_0 + \Delta_N(\phi)\,, \tag{8.10}$$

where the frequency shift $\Delta_N(\phi)$ depends on the CEO phase ϕ and on the harmonic order N. For the special case of $\Delta_N(\phi) = \omega_0$, one obviously gets peaks at the spectral positions of even harmonics (also see Sect. 3.4). This behavior (8.10) immediately reminds us of the equidistant frequency comb of a mode-locked laser oscillator (see Sect. 2.3) that can be shifted by the CEO frequency f_ϕ according to (2.31). In the time domain, the shift $\Delta_N(\phi)$ means that the subsequent attosecond pulses are not identical (as, e.g., in Fig. 5.1) but the phase of the attosecond carrier oscillation with respect to the attosecond envelope, say ϕ_{as}, changes from one attosecond pulse to the next – in strict analogy to what we have discussed for laser pulses in Sect. 2.3 (compare with Figs. 2.2(a) and (b)). This is equivalent to saying that the phases of the individual high harmonics generated in one half of an optical cycle are different from those generated in the next half of the optical cycle. Given the fact that the entire conditions change dynamically during ionization and propagation – depending on ϕ – this might be expected intuitively. In any case, this behavior has been predicted on the basis of numerical solutions of the atom ionization (see Sect. 5.3) using Ammosov–Delone–Krainov ionization rates (see Problem 5.1) coupled to a three-dimensional wave-propagation model in the paraxial approximation [255]. It is also reproduced by the simulations shown in Ref. [254], which account for propagation effects as well.

More recently, CEO-phase effects have also been reported using multicycle optical pulses [256], which are not stabilized in terms of their CEO phase, in a single-shot mode. Here, an interference occurs, e.g., around harmonic order 20 in the wings of the 19th and the 21st harmonic for argon excited by 20-fs pulses at $\hbar\omega_0 = 1.55\,eV$. Also, Rydberg atoms excited by radio-frequency pulses [257] have recently shown a CEO-phase dependence.

---

[1] The authors of Ref. [254] avoid calling the peaks in Fig. 8.4 *high harmonics*. Note, however, that according to the definition of a harmonic of order N used in this book (see Sect. 2.4) they are high harmonics indeed. This does not necessarily mean that they occur at any particular spectrometer frequency ω. Also, see the corresponding discussion in Sect. 3.4 on "third-harmonic generation in the disguise of second-harmonic generation".

8.2 Relativistic Nonlinear Thomson Scattering

In Sect. 8.1.3, we have seen that the *linear* optical contribution of free electrons originating from the ionization of atoms by the laser field can significantly influence the phase matching of high harmonics. Still, at the intensities of $I < 10^{16}\,\text{W/cm}^2$ discussed there, the normalized field strength according to (4.44) follows $|\mathcal{E}| < 0.1$ and *nonlinearities* due to these free electrons from relativistic nonlinear Thomson scattering discussed in Sect. 4.4 are negligible. For light intensities yet one or two orders of magnitude larger, the situation reverses and the nonlinear response is dominated by these free electrons in vacuum. From (4.44) and (2.16) and by inserting the fundamental constants we get the convenient form

$$|\mathcal{E}| = 8.55 \times 10^{-10}\,\frac{\lambda}{\mu\text{m}}\,\sqrt{\frac{I}{\text{W/cm}^2}}\,, \tag{8.11}$$

which directly relates the field parameter to the intensity in units of W/cm^2. λ is the vacuum wavelength corresponding to the laser carrier frequency ω_0 and approximately equals the laser center wavelength for pulses containing many cycles of light. Ref. [258] argues that the remaining Coulomb attraction to the positively charged ions can be neglected.

Experiment

While early experimental work on the interaction of free electrons with light has found harmonics including SHG [259–261], the characteristic emission pattern for each harmonic order $N = 1, 2, 3$ of relativistic nonlinear Thomson scattering (see discussion in Sect. 4.4.1) has first been observed experimentally in Ref. [110]. Note that the radiation pattern depicted for the nonrelativistic regime and $N = 2$ in Fig. 4.5(b) is distorted for relativistic electron motion. Qualitatively, the radiation pattern is stretched along the electron-drift direction parallel to the wavevector of light (analogous to synchrotron emission) and, in addition, the cylindrical symmetry around the wavevector of light is broken, because the electric-field vector accentuates the x-direction. Furthermore, note that in Sect. 4.4 we have discussed the nonlinear emission from *single electrons* only. For a homogeneous gas of electrons excited by a plane electromagnetic wave, phase matching (see Sect. 6.2) is usually not fulfilled and the harmonic intensity is generally strictly zero. In a finite geometry like a laser focus, however, a small but finite contribution remains. The shape of its radiation pattern corresponds to that of the single-electron emission and is sometimes referred to as the incoherent background. Figure 8.5 shows results obtained on a helium jet. The pedestal of the pulse already completely ionizes the gas such that the main part of the pulse interacts predominantly with free electrons. Each dot in the polar diagram corresponds to a single pulse. The mere occurrence of second-harmonic generation is evidence for nonlinear Thomson scattering, the dependence on the azimuthal angle φ is a consequence of relativistic effects.

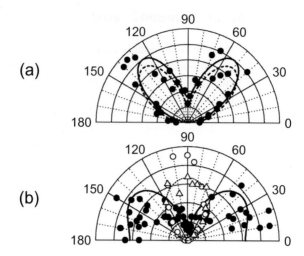

Fig. 8.5. Relativistic nonlinear Thomson scattering on free electrons. Polar diagrams showing the variation of second-harmonic generation ($N = 2$) versus azimuthal angle φ (the angle between the x-axis and the projection of the detection direction onto the xy-plane). Remember that, according to the nomenclature used in this book, the laser electric-field vector $\boldsymbol{E}$ is polarized along x and that the incident wavevector of light $\boldsymbol{K}$ is along z, as illustrated in Fig. 4.5. The angle included by $\boldsymbol{K}$ and the detection direction, θ, equals **(a)** 90 degrees and **(b)** 129 degrees. The full symbols correspond to the experiment, the curves to theory (along the lines of Sect. 4.4.2). The open symbols and the dashed curve in (b) correspond to the fundamental ($N = 1$), which can be compared with Fig. 4.5(a). The experimental parameters are $I = 3.5 \times 10^{18}$ W/cm^2, $\hbar\omega_0 = 1.2$ eV (1.053 μm wavelength), hence $|\mathcal{E}| = 1.5$ [110], 400 fs pulse duration (0.8 J pulse energy) from a Nd:glass laser system and an electron density of $N_e/V = 6.2 \times 10^{19}$ /cm^3. Reprinted by permission from Nature [110] copyright (1998) Macmillan Publishers Ltd.

In contrast to this incoherent background, phase-matched *conical* third-harmonic generation from relativistic electrons has also been observed experimentally [262] (also see Sect. 7.5).

More recently, many even harmonics up to order $N = 30$ [263] at $I = 1 \times 10^{19}$ W/cm^2 have been observed. At yet larger intensities and using a forward scattering geometry, broad continuous spectra have been measured [264] (compare Fig. 4.8 for backscattering). In the latter case, 30-fs pulses with 1.5 J energy from a Ti:sapphire laser system have been focussed on the front of a supersonic helium gas jet to 6-μm spot diameter, leading to an estimated normalized field strength of $|\mathcal{E}| = 5.6$ or $I = 0.7 \times 10^{20}$ W/cm^2. The actual intensities are possibly yet larger ($|\mathcal{E}| \approx 10$) due to relativistic self-focusing. The electron densities are estimated to be in the range from $N_e/V = 10^{18}$ /cm^3 to a few 10^{19} /cm^3. Under these conditions, the authors have observed a peak emission around 150 eV photon energy and a significant contribution in the tail of the spectrum up to $\hbar\omega = 2$ keV.

Theory

Even attosecond pulses have been predicted on different levels of sophistication. Lee et al. [258] suggests attosecond X-ray pulses for $|\mathcal{E}| \approx 10$ on the basis of numerical solutions of the relativistic version of Newton's law (4.40) for optical pulses of finite duration. Naumova et al. [265] performed particle-in-cell simulations that self-consistently solve Maxwell's equations and the relativistic equations of motion of electrons and ions for $|\mathcal{E}| \approx 3$. In particular, they predict efficient ($\approx 10\%$) generation of single attosecond pulses from the mJ to the Joule level.

Let us finally note that our discussion starting from (4.56) – (4.58) up to this point is based on the assumption of *zero* initial electron velocity. The corresponding more general expressions for finite or even relativistic initial electron velocity can, e.g., be found in Ref. [113]. This case, which corresponds to Compton scattering, offers the interesting possibility to transfer momentum and energy from the electron towards the emitted photons. Using "moderate" laser intensities corresponding to $|\mathcal{E}| = 1$, a counterpropagating beam of relativistic electrons and the backscattering geometry, large X-ray powers can possibly be generated [112].

Compact sources of extreme UV or soft X-ray radiation from atoms or from free electrons might find applications in nanometer lithography, in femtosecond X-ray crystallography [266], or in high-resolution imaging.

Solutions

Problems of Chapter 1

1.1 In a "dark" room held at room temperature, one still has the unavoidable blackbody radiation. The corresponding intensity I of electromagnetic radiation is given by Planck's law

$$I = \int_0^\infty \frac{h}{c_0^2} \frac{f^3}{e^{hf/k_B T} - 1} \, df = \frac{h}{c_0^2} \left(\frac{k_B T}{h} \right)^4 \frac{\pi^4}{15} .$$

For $T = 300\,\text{K}$ one gets $I = 0.7 \times 10^{-2}\,\text{W/cm}^2$. Only a very (!) small fraction of this intensity, however, is visible light (photon energies hf in the interval 1.5–3.0 eV). To roughly estimate this fraction, we remember that most of the light stems from the long-wavelength end of the visible spectrum, let us say from the interval 1.5–1.6 eV. The center frequency of that interval is $f = 3.7 \times 10^{14}\,\text{Hz}$, its width is $df = 2.4 \times 10^{13}\,\text{Hz}$. This leads to an estimated intensity of visible light at $T = 300\,\text{K}$ of $I = 10^{-23}\,\text{W/cm}^2$. This value corresponds to a flux of about one photon through your finger tip in six hours time – it does not get any darker at room temperature.

Note that it takes 22 orders of magnitude to get from the intensity of a "dark" room to that of the sun on the earth. We will go another 29 orders of magnitude upwards in this book.

Problems of Chapter 2

2.1 Under these conditions, the intensity $I \propto \sqrt{\epsilon} \, \tilde{E}_0^2$ (see (2.16)) is the same in air and in the dielectric. In air $\epsilon = 1$, in the dielectric $\epsilon = 10.9$. $\Rightarrow \tilde{E}_0$ in the dielectric is given by $4 \times 10^9\,\text{V/m}/\sqrt{\sqrt{10.9}} = 2.2 \times 10^9\,\text{V/m}$.

2.2 There is no unique answer to this imprecise question ... The most stringent way to interpret the question is that all frequency components of the pulse must lie in the visible, i.e., roughly in the photon energy interval 1.5–3.0 eV, one octave. Let us assume that all frequency components have the same amplitude and the same phase, which leads to the shortest pulse. This "box-spectrum" with a FWHM of $\hbar\delta\omega = 1.5\,eV$, $\Leftrightarrow \delta\omega = 2.28\,/fs$, leads to $\mathrm{sinc}^2(t)$–pulses of $\delta t = 2\pi \times 0.8859/\delta\omega = 2.44\,fs$ in duration (for the duration–bandwidth product $\delta\omega\,\delta t$ see footnote in Sect. 2.3). With a carrier frequency of $\hbar\omega_0 = 2.25\,eV$, $\Leftrightarrow 2\pi/\omega_0 = 1.84\,fs$ period of light, these pulses contain 1.3 optical cycles. The electric field of these pulses with $\delta\omega/\omega_0 = 2/3$ is very nearly similar to that depicted in Fig. 2.3, where $\delta\omega/\omega_0 = 0.60$ has been chosen.

By the way: The number of locked modes is given by the width of the spectrum devided by the distance between adjacent modes, i.e., by the ratio $\delta\omega/\Delta\omega$. With $\Delta\omega = 2\pi \times 100\,MHz$ from Example 2.2 and $\delta\omega = 2.28\,/fs$ this leads to 3.6×10^6 locked modes.

2.3 We want to compute the Fourier transform (FT) of the N-th harmonic from a $\chi^{(N)}$ process, which is proportional to the FT of $(\tilde{E}(t))^N \cos(N\omega_0 t + N\phi)$. The envelope $\tilde{E}(t)$ ought to be "well behaved". We specify that it may contain maxima/minima of *different* heights, the largest maximum of which shall be centered around $t = 0$. The key is to realize that the N-th power of $\tilde{E}(t)$ highlights a narrow time window around $t = 0$. If, for example, the envelope has decreased to 90% of the peak, the 51st power has already dropped to the negligible amount of $0.9^{51} = 0.0046$. This allows for a truncated Taylor expansion of the envelope around $t = 0$, i.e.,

$$\left(\tilde{E}(t)\right)^N = \left(\tilde{E}_0 + \underbrace{\frac{1}{1!}\frac{d\tilde{E}}{dt}(0)\,t}_{=\,0} + \underbrace{\frac{1}{2!}\frac{d^2\tilde{E}}{dt^2}(0)}_{=:\,-\mathcal{C}\tilde{E}_0\,\leq\,0}\,t^2 + \ldots\right)^N$$

$$\approx \left(\tilde{E}_0\left(1 - \mathcal{C}t^2\right)\right)^N$$

$$= \tilde{E}_0^N \sum_{n=0}^{N} \underbrace{\binom{N}{n}}_{\approx\,\frac{N^n}{n!}\ \text{for}\ n \ll N} \left(-\mathcal{C}t^2\right)^n$$

$$\approx \tilde{E}_0^N \sum_{n=0}^{\infty} \frac{1}{n!}\left(-N\mathcal{C}t^2\right)^n \quad \text{for}\ N \to \infty$$

$$= \tilde{E}_0^N\,e^{-N\mathcal{C}t^2}.$$

The temporal width of this Gaussian obviously scales $\propto 1/\sqrt{N\mathcal{C}}$ and is solely determined by the parameter $\mathcal{C} \geq 0$, which is proportional to the *curvature* of the envelope at the maximum. The FT of this Gaussian is also Gaussian with width $\propto \sqrt{N\mathcal{C}}$. Note that our reasoning fails for box-shaped[1] or flat-top pulse envelopes where $\mathcal{C} = 0$.

---

[1] The N-th power of a box-shaped pulse is again box-shaped for arbitrary N. Thus, all harmonics strictly have the identical sinc^2-shape in the intensity spectrum.

Hence, we have to add $C \neq 0$ to our above checklist for "well behaved" pulses. For example, sinc, Lorentzian or sech-envelopes $\tilde{E}(t)$ with/without additional temporal satellite pulses of whatever shape would fall into the category "well behaved".

Alternatively, one can argue within the frequency domain to obtain the same result: The N-th harmonic spectrum is proportional to the N-fold convolution of the laser spectrum with itself, which, for large N, again leads to a Gaussian shape.

2.4 We start from a sequence of pulses (not necessarily a periodic train of pulses), the CEO phase ϕ of which might fluctuate randomly. The pulse envelope and hence the spectrum, on the other hand, are assumed to be stable. This sequence of pulses could also be replaced by a sequence of pulses that has previously been spectrally broadened by SPM – which does not change their CEO phase. The idea: *If we could take advantage of a difference-frequency mixing process with phase $\phi - \phi = 0$, the phase would obviously drop out.* In order to get there, let us filter out two contributions from the original spectrum, one at the low-frequency end, with carrier frequency ω_1, and another one at the high-frequency end, with carrier frequency ω_2. They correspond to temporal pulse envelopes $\tilde{E}_1(t)$ and $\tilde{E}_2(t)$, respectively. Each of these pulses is clearly longer than the original one, but they both have the same CEO phase ϕ (or, depending on the choice of the carrier frequencies ω_1 and ω_2, a CEO phase given by ϕ plus some constant offset – which is omitted for clarity). Sending the two pulses onto a $\chi^{(2)}$ medium generates the electric field $E^{(2)}(t)$ given by (see (2.39))

$$E^{(2)}(t) \propto \chi^{(2)} \frac{\partial^2}{\partial t^2} \left(\tilde{E}_1(t) \cos(\omega_1 t + \phi) + \tilde{E}_2(t) \cos(\omega_2 t + \phi) \right)^2$$

$$= \chi^{(2)} \frac{\partial^2}{\partial t^2} \left(\tilde{E}_1(t) \tilde{E}_2(t) \cos((\omega_2 - \omega_1) t + (\phi - \phi)) + \ldots \right)$$

$$\approx -\chi^{(2)} \omega_{\mathrm{DFG}}^2 \tilde{E}_1(t) \tilde{E}_2(t) \cos(\omega_{\mathrm{DFG}} t + 0) + \ldots$$

$$= \tilde{E}_{\mathrm{DFG}}(t) \cos(\omega_{\mathrm{DFG}} t + 0) + \ldots .$$

Here we have again ignored propagation effects within the nonlinear medium and we have omitted all terms except for the difference-frequency generation (DFG) at carrier frequency $\omega_{\mathrm{DFG}} = (\omega_2 - \omega_1)$ from the second line on. Towards the third line we have neglected the temporal derivatives of the envelopes. In the fourth line we have lumped all prefactors into an effective DFG envelope $\tilde{E}_{\mathrm{DFG}}(t)$. Obviously, the phase of the resulting pulse is zero, hence *constant* – even for fluctuating ϕ. This contribution can be separated from SHG and sum-frequency generation via its wavevector of light. If the original pulse spans more than one octave, i.e., if $\omega_2 > 2\omega_1$, the difference frequency $(\omega_2 - \omega_1) > \omega_1$ even lies within the original spectrum.

A corresponding scheme using a seeded optical parametric amplifier has been discussed in Ref. [38]. This scheme is useful for amplified laser pulses. Experiments using pulses directly from a laser oscillator, spectrally broadened via self-phase modulation, have also been reported [39].

2.5 The laser pulse $E_{\phi=0}(t)$ in the time domain corresponds to a spectrum $E_{\phi=0}(\omega)$ in the Fourier domain (also see Example 2.4). We can take advantage of the fact that the CEO phase ϕ corresponds to a phase factor in the spectral domain, i.e.,

$$E_{\phi=0}(\omega) \rightarrow E_{\phi\neq0}(\omega) = E_{\phi=0}(\omega)\, e^{-i\phi} \quad \text{for } \omega \geq 0,$$

and

$$E_{\phi=0}(\omega) \rightarrow E_{\phi\neq0}(\omega) = E_{\phi=0}(\omega)\, e^{+i\phi} \quad \text{for } \omega \leq 0.$$

Transforming back into the time domain delivers the desired laser electric field $E_{\phi\neq0}(t)$. Note that this procedure requires just two (numerical) Fourier transformations and does not explicitly decompose the pulse into an envelope and a carrier-wave oscillation. We do not even have to specify the carrier frequency ω_0. This procedure does, however, tacitly assume that the positive and negative frequency components do not overlap. This condition is usually fulfilled even for single-cycle pulses (see Example 2.4). A notable exception are the box-shaped pulses discussed in Sect. 3.5.

2.6 Within the narrow-band or long-pulse limit, the shape as well as the temporal duration of the pulse, Δt, remain unchanged when going from vacuum into the dielectric. For a plane wave propagating along the z-direction, the duration is connected to the longitudinal extent of the pulse envelope in vacuum and in the dielectric medium via $\Delta z_{\text{vac}} = c_0\,\Delta t$ and $\Delta z_{\text{med}} = v_{\text{group}}\,\Delta t$, respectively. Thus, within the dielectric, the factor $v_{\text{group}}/c_0 \leq 1$ compresses the pulse in real space and its energy density increases: As usual, the electromagnetic energy per volume is given by $\frac{1}{2}(\boldsymbol{D} \cdot \boldsymbol{E} + \boldsymbol{B} \cdot \boldsymbol{H}) = \epsilon_0 \epsilon E^2$. If we neglect absorption and reflection of electromagnetic energy at the air/dielectric interface, the total energy is the same if the pulse is either entirely in vacuum or entirely in the dielectric. Thus, we have

$$\epsilon_0 \tilde{E}_{0,\,\text{vac}}^2\, \Delta z_{\text{vac}} = \epsilon_0 \epsilon \tilde{E}_{0,\,\text{med}}^2\, \Delta z_{\text{med}}$$

$$= \epsilon_0 \tilde{E}_{0,\,\text{vac}}^2\, c_0\, \Delta t = \epsilon_0 \epsilon \tilde{E}_{0,\,\text{med}}^2\, v_{\text{group}}\, \Delta t\,,$$

where $\tilde{E}_{0,\,\text{vac}}$ and $\tilde{E}_{0,\,\text{med}}$ are the peak of the field envelope in vacuum and in the dielectric medium, respectively. Solving for $\tilde{E}_{0,\,\text{med}}^2$ we get

$$\tilde{E}_{0,\,\text{med}}^2 = \frac{1}{\epsilon}\,\frac{c_0}{v_{\text{group}}}\,\tilde{E}_{0,\,\text{vac}}^2\,.$$

For the corresponding peak intensities $I \propto \sqrt{\epsilon}\,\tilde{E}_0^2$ according to (2.16) this leads to

$$I_{\text{med}} = \frac{1}{\sqrt{\epsilon}}\,\frac{c_0}{v_{\text{group}}}\,I_{\text{vac}} = \frac{v_{\text{phase}}}{v_{\text{group}}}\,I_{\text{vac}}\,.$$

It is straightforward to additionally account for reflection losses, which, in the narrow-band limit, only depend on $\epsilon(\omega_0)$ via the Fresnel coefficients (and not on v_{group}). For "slow light", i.e., for $v_{\text{group}} \rightarrow 0$, the peak electric field and the intensity within the dielectric can become much larger than in vacuum, enhancing the effective optical nonlinearities. For the special case $v_{\text{group}} = v_{\text{phase}} = c_0/\sqrt{\epsilon}$, we recover $I_{\text{med}} = I_{\text{vac}}$ and $\tilde{E}_{0,\,\text{med}} = \tilde{E}_{0,\,\text{vac}}/\sqrt{\sqrt{\epsilon}}$ (see Problem 2.1).

Problems of Chapter 3

3.1 Introducing a Stokes damping γ into Newton's second law (3.1) leads to the real part of the refractive index

$$n(\omega) = \mathrm{Re}\left(\sqrt{1 + \chi(\omega)}\right),$$

with the susceptibility

$$\chi(\omega) = \underbrace{\frac{e^2 N_{\mathrm{osc}}}{\epsilon_0 V m_e}}_{:= \omega_{\mathrm{pl}}^2} \frac{1}{\Omega^2 - \omega^2 - i\gamma\omega}$$

(compare (3.3)). The absorption coefficient is

$$\alpha(\omega) = 2\frac{\omega}{c_0} \mathrm{Im}\left(\sqrt{1 + \chi(\omega)}\right).$$

As usual, phase and group velocities are given by

$$v_{\mathrm{phase}}(\omega) = \frac{\omega}{K} = \frac{\omega}{\frac{\omega}{c_0} n(\omega)} = \frac{c_0}{n(\omega)}$$

and

$$v_{\mathrm{group}}(\omega) = \frac{d\omega}{dK} = \frac{1}{\frac{d}{d\omega}K} = \frac{1}{\frac{d}{d\omega}\left(\frac{\omega}{c_0} n(\omega)\right)} = \frac{c_0}{n(\omega) + \omega \frac{dn}{d\omega}}$$

via the frequency ω and the real part of the wave number K. The group velocity shown in Fig. 3.16 exhibits a rather complex behavior, whereas the index profile $n(\omega)$ is as usual. Close to resonance, $\omega \approx \Omega$, we have anomalous dispersion with $\frac{dn}{d\omega} < 0$. This reduces the denominator in the expression for $v_{\mathrm{group}}(\omega)$ below unity and leads to superluminal group velocities. For large negative slopes, even the group velocity itself becomes negative. Let us focus our discussion on frequencies ω with $v_{\mathrm{group}}(\omega) > c_0$ or $v_{\mathrm{group}}(\omega) < 0$ (dark gray area in Fig. 3.16). Be aware that this spectral region experiences strong absorption. For example, under the conditions of Fig. 3.16 and for $\hbar\Omega = 1.5\,eV$, the absorption length would be merely $0.2\,\mu m$. For a *long Gaussian pulse* whose spectrum is narrow and centered in this region, and for a medium that is only few absorption lengths in thickness, the pulse actually seems to travel with a speed exceeding c_0, while approximately maintaining its shape (we give an intuitive explanation below) – but the amplitude of the pulse decays exponentially during propagation. This is the Garrett and Mc Cumber effect [41,42] that has been observed experimentally [43]. For a *short Gaussian pulse* whose spectrum encompasses the resonance, some frequency components propagate with superluminal velocity. Other parts of the spectrum propagate with luminal or subluminal velocity and with much less attenuation. Thus, one obtains substantial *spectral reshaping* of the pulse. Under

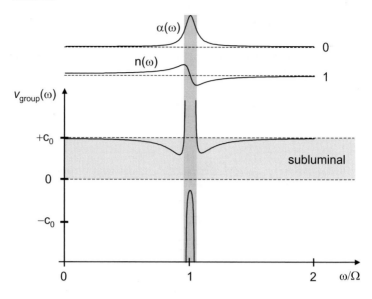

Fig. 3.16. Group velocity $v_{group}(\omega)$ of the Lorentz oscillator model. The upper two curves show the absorption spectrum $\alpha(\omega)$ and the refractive index $n(\omega)$ spectrum, respectively. The dashed horizontal lines correspond to $\alpha = 0$ and $n = 1$, respectively. Parameters are: $\gamma/\Omega = 0.1$ and $\omega_{pl}^2/\Omega^2 = 0.05$. With decreasing damping γ/Ω, the dark gray region (corresponding to anomalous dispersion) shrinks in width.

these broadband conditions, the group velocity at a particular frequency – for instance at the carrier frequency $v_{group}(\omega_0)$ – loses its meaning. Remember that the concept of the group velocity is usually introduced by a Taylor expansion of the dispersion relation $K(\omega)$ or by considering the beating of two nearby frequency components.

Under broadband Gaussian conditions, the *pulse centrovelocity* v_{centro} [44] is a helpful extension. It is related to the average *energy flow* of an optical pulse in a dispersive medium and is distinct from the energy transport velocity [45]. For a plane wave with a wavevector K directed along the z-coordinate and with E and H polarized along x and y, respectively, we define

$$v_{centro} = \frac{z}{\bar{t}(z)},$$

where the "center of mass" time $\bar{t}(z)$ for propagation from 0 (e.g., the front of a sample) towards coordinate z (e.g., the end of a sample) is defined as

$$\bar{t}(z) = \frac{\displaystyle\int_{-\infty}^{+\infty} t\, S(z, t)\, dt}{\displaystyle\int_{-\infty}^{+\infty} S(z, t)\, dt}.$$

$S(z, t)$ is the z-component of the Poynting vector. Note that the pulse centrovelocity refers to a possible measurement. It can be shown [44] that the delay $\bar{t}(z)$ consists of two contributions: A *net group delay*, essentially the average of the group delay $z/v_{\text{group}}(\omega)$ weighted with the pulse spectrum, and a so-called *reshaping delay* arising from the frequency-dependent transmission. According to our above discussion, the reshaping delay vanishes for spectrally narrow pulses and samples with a thickness of only a few absorption lengths, thus $v_{\text{centro}} = v_{\text{group}}(\omega_0)$. On the other hand, for broadband pulses, $v_{\text{centro}} \neq v_{\text{group}}(\omega_0)$. A corresponding discussion, exemplified on the Lorentz oscillator model, can be found in Ref. [44].

Let us now give the promised intuitive explanation for the Gaussian narrow-band, thin-sample limit for which the time delay $\bar{t}(z)$, the pulse centrovelocity, and the group velocity can all become negative (see numerically exact solutions shown in Fig. 3.17). According to the above definition, a negative time delay $\bar{t}(z)$ occurs when the "center of mass" of the transmitted pulse emerges from the sample at an instant before the

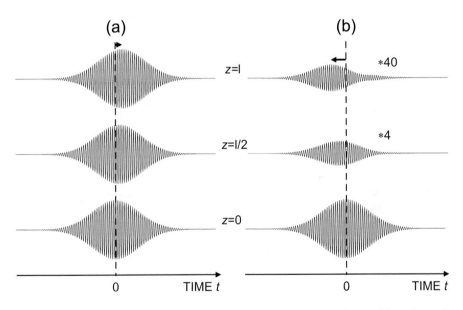

Fig. 3.17. Electric field $E(z, t)$ versus time t at three positions: $z = 0$, $z = l/2$, and $z = l$. **(a)** Propagation in vacuum, i.e., $v_{\text{group}}(\omega_0) = c_0 = v_{\text{phase}}(\omega_0)$, for reference. **(b)** Propagation through a resonant Lorentz oscillator medium under conditions of negative group velocity, i.e., $v_{\text{group}}(\omega_0) < 0$, and also $v_{\text{centro}} < 0$. The electric field is obtained numerically via $E(z, t) \propto \int_0^\infty E_+(0, \omega) \exp(i\frac{\omega}{c_0}\sqrt{1 + \chi(\omega)}\, z - i\omega t)\, d\omega + \text{c.c.}$ Here, $E_+(0, \omega)$ is the positive-frequency part of the spectrum of the Gaussian pulse at $z = 0$ given by $E_+(0, \omega) \propto \exp(-(\omega - \omega_0)^2/\sigma^2)$ (see Example 2.4). The oscillator parameters are: $\omega_{\text{pl}}^2/\Omega^2 = 0.05$ and $\gamma/\Omega = 0.1$ (see Fig. 3.16). The pulse parameters are: $\omega_0/\Omega = 1$ and $\sigma/\omega_0 = 0.02$. The medium has a thickness of $l = 20\, c_0/\omega_0$. Under the above conditions, this corresponds to about 12 absorption lengths, i.e., $\alpha(\omega_0)\, l \approx 12$. At this point, one can already see distortions of the Gaussian pulse shape. The curves in (b) are multiplied by the indicated factors to compensate for the exponential attenuation.

"center of mass" of the incident pulse enters the sample. This *appears* to violate causality, even though the result is based on the (causal) Maxwell equations. Let us discuss this situation in the time domain: First, remember that absorption occurs if the oscillator follows the excitation with a $\pi/2$ phase shift. As a resonant pulse with $\omega_0 \approx \Omega$ approaches the sample surface, its leading edge excites oscillators in the medium. On a timescale roughly determined by the inverse damping rate γ, the oscillators have a phase shift different from $\pi/2$ and some light can pass the sample. For longer times approaching the "center of mass" of the incident pulse, a $\pi/2$ phase shift evolves and light is heavily absorbed. This effectively "chops off" the center as well as the trailing edge of the incident pulse. The "center of mass" of the emerging pulse stems from the leading edge of the incoming pulse. All the rest is absorbed. There is no problem with causality. The fact that the pulse shape does not appear to be distorted – even though it is strongly distorted – is a particularity of Gaussian pulses (or pulses close to a Gaussian) in the long-pulse, thin-sample limit [41].

This overall reasoning is justified for samples with a thickness of ten absorption lengths or less [41, 43]. For many more absorption lengths, sometimes called the asymptotic ($z \to \infty$) description or the "mature dispersion limit", the oscillators at the end of the sample no longer experience the incident field but rather a strongly modified driving field. Be aware that even a very narrow incident Gaussian spectrum $E_+(0, \omega) \propto \exp(-(\omega - \omega_0)^2/\sigma^2)$ with $\omega_0 = \Omega$ and $\sigma \ll \gamma$ at $z = 0$ can experience a huge ($\gg \sigma$) spectral redshift of its intensity spectrum $\propto |E_+(z, \omega)|^2 = |E_+(0, \omega)|^2 \exp(-\alpha(\omega) z)$ if z corresponds to hundreds or thousands of absorption lengths. If one starts on the high-energy wing of the absorption line, one gets a huge blueshift. Thus all of the relevant frequency components of the pulse eventually ($z \to \infty$) lie in the region of *subluminal* group velocity where the absorption is lower and the pulse propagates slower than c_0. Readers who wish to speculate about implications on the speed of information transfer over "long" distances should keep this in mind. In any case, the light emerging from the sample is attenuated by tens of orders of magnitude. The asymptotic limit $z \to \infty$ has been discussed for the narrow-band Gaussian case [46], for box-shaped pulses [47], and for pulses with a steep rising edge [48]. The "mature" spectral chirp acquired during propagation over such long distances leads to the Sommerfeld and the Brillouin precursor [45, 49] in the time domain. By the way: The very intuitive acoustic analogue is discussed in Ref. [50].

3.2 With the definition of the dipole matrix element

$$d = \int_{-\infty}^{+\infty} \psi_2^*(x)(-e\,x)\,\psi_1(x)\,dx ,$$

and with the wave functions

$$\psi_1(x) = \sqrt{\frac{2}{L}} \sin\left(\frac{1\,\pi}{L}\,x\right)$$

$$\psi_2(x) = \sqrt{\frac{2}{L}} \sin\left(\frac{2\,\pi}{L}\,x\right)$$

we obtain

$$d = -\frac{2e}{L} \int_0^L \sin\left(\frac{2\pi}{L}x\right) x \sin\left(\frac{\pi}{L}x\right) dx$$

$$= -\frac{2e}{L}\frac{L^2}{\pi^2} \underbrace{\int_0^\pi \sin(2X)\, X \sin(X)\, dX}_{=-8/9}$$

$$= \frac{16}{9\pi^2}\, eL \approx 18\%\, eL\,.$$

This is no surprise – the dipole matrix element is proportional to the width of the box L. Also, it can already be seen from Fig. 3.2 that d has to be substantially smaller than eL because the excursion of the electron center of mass is much smaller than L.

Similarly, the dipole moment d for a transition from the ground state to an arbitrary state number N in the well is zero for all odd integer N. We obtain, e.g., $d \approx 1.4\%\, eL$ for $N = 4$ and $d \approx 0.4\%\, eL$ for $N = 6$. Obviously, the transition from the ground state to the first excited state has a much larger dipole moment than all other transitions. This a posteriori justifies the two-level system approximation for this model problem.

3.3 The complete optical Bloch equations including dephasing read

$$\dot{u} = +\Omega v - u/T_2$$
$$\dot{v} = -\Omega u - 2\Omega_R w - v/T_2$$
$$\dot{w} = +2\Omega_R v\,.$$

In the stationary limit we have $\dot{w} = 0$. From the third line it follows that $v = 0$, hence $\dot{v} = 0$. Inserting $v = 0$ into the first line we, furthermore, get $u = 0$. From the second line with $\dot{v} = u = v = 0$ we finally get $w = f_2 - f_1 = 0$, i.e., with $f_2 + f_1 = 1$, the occupation of the excited state and of the ground state are both 50%, the system is transparent. Note that the only point at which the strong dephasing assumption has entered is the existence of a stationary limit. According to the equations of motion for u and v, this limit is reached on a timescale comparable to T_2.

If the inversion additionally experiences relaxation towards the ground state, i.e., towards $w = -1$, according to $\dot{w} = -(w + 1)/T_1$, with the occupation relaxation time or longitudinal relaxation time T_1, the steady-state inversion is generally smaller than zero, i.e., $-1 \le w \le 0$. In the limit $T_1 \to 0$, we get $w = -1$ ($\Leftrightarrow f_2 = 0$ and $f_1 = 1$) from the third line of the optical Bloch equations, i.e., from $\dot{w} = 0 = +2\Omega_R v - (w + 1)/T_1 \approx -(w + 1)/T_1$.

3.4 From Fig. 3.1 we see that the positive frequency part of the laser spectrum resonantly ($\omega_0 = \Omega$) excites the positive frequency pole of the two-level system resonance indeed. The maximum of the laser spectrum at negative frequencies, on the other hand, has a detuning of $2\omega_0 = 2\Omega$. This leads to a beat note with that detuning, which is the origin of the rapidly oscillating component of the inversion $w(t)$ in Figs. 3.4(a) and (b).

3.5 In order to obtain a nonvanishing *"THG in the disguise of SHG"* contribution, the low-frequency end of the THG spectrum needs to overlap with the spectrometer frequency $2\omega_0$. For sinc^2-pulses and within the $\chi^{(3)}$ limit, the width of the THG spectrum with carrier frequency $3\omega_0$ is three times that of the fundamental spectrum, i.e., it is given by $3\,\delta\omega$. This leads to the condition $2\omega_0 = 3\omega_0 - 3\,\delta\omega/2 \Leftrightarrow \delta\omega/\omega_0 = 2/3$, equivalent to a spectral width of one octave (see solution of Problem 2.2). With the duration–bandwidth product of $\delta\omega\,\delta t = 2\pi \times 0.8859$ for an unchirped sinc^2-pulse (see footnote in Sect. 2.3 or solution of Problem 2.2) and with $\hbar\omega_0 = 1.5\,eV$, this corresponds to a maximum pulse width of $t_{FWHM} = \delta t = 3.6\,fs$. Note, however, that the *"THG in the disguise of SHG"* signal would still be arbitrarily small at this point. Thus, the 5-fs pulses discussed in Sect. 7.2 can only lead to significant *"THG in the disguise of SHG"* deep inside the nonperturbative regime.

3.6 In order to arrive at an instantaneous response according to (2.37), the two-level system must clearly react instantaneously. In other words: The time derivative of the Bloch vector must be approximately zero (adiabatic following). For a harmonic oscillator, this is fulfilled if the transition frequency Ω is much larger than the driving frequency ω_0, i.e. for $\Omega/\omega_0 \gg 1$. The inversion also follows adiabatically if the longitudinal damping $1/T_1$ is large. In these limits, the Bloch equations (also see Problem 3.3) become

$$0 = +\Omega v - u/T_2$$
$$0 = -\Omega u - 2\,\Omega_R(t)w - v/T_2$$
$$0 = +2\,\Omega_R(t)v - (w+1)/T_1\,.$$

Solving the first line for u, inserting this result into the second line, solving for v and inserting into the third line delivers

$$w(t) = -\cfrac{1}{1 + \cfrac{T_1/T_2}{\Omega^2 + 1/T_2^2}\,4\Omega_R^2(t)}$$

and

$$u(t) = -\frac{2\,\Omega_R(t)\,\Omega}{\Omega^2 + 1/T_2^2}\,w(t)\,.$$

In the perturbative limit, i.e., for $\Omega_R/\Omega \ll 1$, we can expand the expression for $w(t)$ in a Taylor series using $1/(1+x) \approx (1-x)$ for $x \ll 1$. With the macroscopic optical polarization P according to (3.20), this immediately leads to

$$P(t) = \frac{N_{2LS}}{V}\,d\,u(t)$$

$$= \frac{N_{2LS}}{V}\,d\,\frac{2\,\Omega_R(t)\,\Omega}{\Omega^2 + 1/T_2^2}\left(1 - \frac{T_1/T_2}{\Omega^2 + 1/T_2^2}\,4\Omega_R^2(t) + \dots\right)$$

$$= \epsilon_0\left(\chi^{(1)}E(t) + \chi^{(3)}E^3(t) + \dots\right)\,.$$

From the second to the third line we have inserted for the Rabi frequency $\hbar\Omega_R(t) = dE(t)$ and lumped all prefactors into $\chi^{(1)}$ and $\chi^{(3)}$. This form is obviously identical to (2.37), with $\chi^{(2)} = 0$ because of inversion symmetry. $\chi^{(3)}$ is negative. Note that, in the limit $\Omega \to \infty$, the linear susceptibility scales as $\chi^{(1)} \propto 1/\Omega$ and the third-order susceptibility according to $\chi^{(3)} \propto 1/\Omega^3$. Also note that $\chi^{(3)} \to 0$ for $T_2 \to \infty$, i.e., if we had not introduced a finite transverse damping $1/T_2$ at the beginning, $\chi^{(3)}$ would have been zero.

We summarize for the odd-order nonlinear optical susceptibilities $\chi^{(N)}$ of the two-level system in the far off-resonant perturbative limit

$$\left|\chi^{(N)}\right| \propto \left(\frac{\Omega}{\omega_0}\right)^{-N} \quad \text{for} \quad \frac{\Omega}{\omega_0} \gg 1 \quad \text{and} \quad \frac{\Omega_R}{\Omega} \ll 1 \, .$$

3.7 Taking the temporal derivative of the first line of the Bloch equations (3.17) and inserting $\dot{v}$ from the second line gives

$$\ddot{u} + \Omega^2 u = -2\Omega\Omega_R w \, .$$

Similarly, we obtain with the third and the second line

$$\ddot{w} + 4\Omega_R^2 w = -2\Omega\Omega_R u \, .$$

The physics is simply that of two coupled harmonic oscillators with "displacements" u and w. Without the coupling on the RHS, one oscillator would have eigenfrequency Ω, the other one $2\Omega_R$. The ansatz that $u(t)$ and $w(t)$ both oscillate as a sine or cosine with frequency Ω_{eff} solves the above equations for

$$\Omega_{\text{eff}} = \sqrt{4\Omega_R^2 + \Omega^2} \, .$$

This is just (3.32). The desired initial condition $(u(t_0), v(t_0), w(t_0))^T$ of the Bloch vector at time $t_0 = 0$ can be fulfilled by a linear combination of these solutions. With some patience, this leads to (3.31).

3.8 The level diagram is schematically shown in Fig. 3.18.

3.9 Introducing the ratio $\mathcal{R} = a_2/a_1$, it is easy to arrive at the Riccati-type equation of motion [84, 85]

$$\frac{d\mathcal{R}}{d(\omega_0 t)} = i\frac{\Omega_R(t)}{\omega_0} - i\frac{\Omega}{\omega_0}\mathcal{R} - i\frac{\Omega_R(t)}{\omega_0}\mathcal{R}^2 \, ,$$

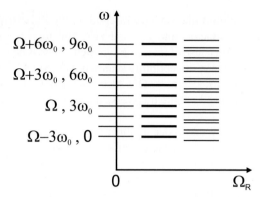

Fig. 3.18. Scheme of the eigenfrequencies ω of the Hamilton matrix (3.55) for off-resonant excitation, i.e., for $\Omega = 3\,\omega_0$. Note that, for a given Rabi frequency Ω_R/Ω, the splitting is much smaller here as compared to the case of resonant excitation, where $\Omega = \omega_0$ (see Fig. 3.15).

with $\mathcal{A}_2(t) = -\mathcal{A}_0(t)$. The inversion w of the two-level system, for example, results as

$$w = |a_2|^2 - |a_1|^2 = \frac{|\mathcal{R}|^2 - 1}{|\mathcal{R}|^2 + 1},$$

where we have used the normalization condition $|a_1|^2 + |a_2|^2 = 1$.

Problems of Chapter 4

4.1 (a) The dipole moment of one oscillator is $(d\,u)$ (see (7.6) and (7.14)) or $(-e\,x)$. The peak value of $u = 1$ is reached for a Rabi oscillation with $\tilde{E}_0 = 4 \times 10^9$ V/m in GaAs, hence, e.g., for $d = 0.5\,e$ nm (GaAs) we obtain $x_0 = 0.5$ nm $\approx a \propto (\tilde{E}_0)^0$.

(b) From Newton's law we get the peak electron displacement

$$x_0 = \frac{e\tilde{E}_0}{m_e\omega_0^2}$$

$\propto (\tilde{E}_0)^{+1}$. For ZnO parameters ($m_e = 0.24 \times m_0$, $\hbar\omega_0 = 1.5\,eV$) we obtain $x_0 = 0.56$ nm $\approx a$.

Even for laser intensities around $I = 2 \times 10^{12}$ W/cm^2, $\Leftrightarrow \tilde{E}_0 = 4 \times 10^9$ V/m (see Example 2.1), the classical crystal-electron displacements stay on the order of one lattice constant a.

4.2 As pointed out above, it is convenient to express the laser electric field as a sine at this point, i.e., $E(t) = \tilde{E}_0 \sin(\omega_0 t + \phi')$. With $\phi' = \phi + \pi/2$, this is equivalent to the form otherwise used in this book, i.e., $E(t) = \tilde{E}_0 \cos(\omega_0 t + \phi)$ with the CEO phase ϕ. The vector potential becomes $A_x(t) = -\tilde{E}_0/\omega_0 \cos(\omega_0 t + \phi')$. Inserting the ansatz (4.18) into the Schrödinger equation in the radiation gauge, and after some straightforward algebra, we obtain the relation for the coefficients a_N

$$2N\, a_N = -\underbrace{\frac{e\tilde{E}_0 k_x}{m_e \omega_0^2}}_{=:\, u} \left(a_{N+1} + a_{N-1}\right) + \underbrace{\frac{\langle E_{\text{kin}} \rangle}{\hbar \omega_0}}_{=:\, 2v} \left(a_{N+2} + a_{N-2}\right).$$

Note that this relation contains real quantities only. Reiss [90] introduced the generalized Bessel functions

$$J_N(u, v) = \sum_{M=-\infty}^{+\infty} J_M(v)\, J_{N-2M}(u)\,,$$

which fulfill the mathematical identity ((B10) in that paper)

$$2N\, J_N(u, v) = u\left(J_{N+1}(u, v) + J_{N-1}(u, v)\right) + 2v\left(J_{N+2}(u, v) + J_{N-2}(u, v)\right).$$

Comparison of the coefficients delivers $a_N = J_N(u, v)$. Switching back to our definition of the laser electric field, we obtain $a_N \to \exp(i\frac{\pi}{2} N)\, a_N = \exp(i\frac{\pi}{2} N)\, J_N(u, v)$, which is simply (4.24).

4.3 Following the lines of Sect. 4.3.2, it is easy to arrive at

$$v_{\text{group}}(t) = -\frac{a\Delta}{\hbar} \sin\left(\frac{\Omega_B}{\omega_0} \sin(\omega_0 t) + \Omega_B^{\text{dc}} t\right).$$

Upon using the identity $\sin(X + Y) = \sin(X)\cos(Y) + \cos(X)\sin(Y)$ we further get

$$
\begin{aligned}
v_{\text{group}}(t) = -\frac{a\Delta}{\hbar} &\left\{ \sin\left(\frac{\Omega_B}{\omega_0} \sin(\omega_0 t)\right) \cos(\Omega_B^{\text{dc}} t) \right.\\
&\left. + \cos\left(\frac{\Omega_B}{\omega_0} \sin(\omega_0 t)\right) \sin(\Omega_B^{\text{dc}} t) \right\}\\
= -\frac{a\Delta}{\hbar} &\left\{ \left[2 \sum_{M=0}^{\infty} J_{2M+1}\left(\frac{\Omega_B}{\omega_0}\right) \sin\big((2M+1)\,\omega_0 t\big)\right] \cos(\Omega_B^{\text{dc}} t) \right.\\
&\left. + \left[J_0\left(\frac{\Omega_B}{\omega_0}\right) + 2 \sum_{M=1}^{\infty} J_{2M}\left(\frac{\Omega_B}{\omega_0}\right) \cos(2M\omega_0 t)\right] \sin(\Omega_B^{\text{dc}} t) \right\}.
\end{aligned}
$$

In the last step, we have employed two of the mathematical identities for Bessel functions (see formulae 9.1.42 and 9.1.43 in Ref. [80]). For $\Omega_B^{\text{dc}} = 0$, we recover (4.38). Let us discuss just one aspect of the case $\Omega_B^{\text{dc}} \neq 0$: Not only odd, but also even, $N = 2M$, harmonics occur in the radiated intensity spectrum $I_{\text{rad}}(\omega) \propto |\omega\, v_{\text{group}}(\omega)|^2$. This effect simply arises from the breaking of the inversion symmetry via the dc field and is fairly robust against damping (scattering), also see [103]. Second-harmonic generation from a biased n-doped GaAs/AlAs superlattice has indeed been observed experimentally at $\hbar\omega_0 = 2.9\,\text{meV}$ [104].

4.4 With Stokes damping and within the nonrelativistic regime, we have the equation of motion for the velocity component β_x

$$\frac{d\beta_x}{d\tilde{t}} + \frac{\beta_x}{\tau} = \mathcal{E}\cos(\tilde{t}).$$

Here we have introduced the (normalized) damping time τ. With the initial condition $\beta_x(0) = \beta_x^0$, the solution is

$$\beta_x(\tilde{t}) = \mathcal{E}\frac{\tau}{1+\tau^2}\left(\cos(\tilde{t}) + \tau\sin(\tilde{t}) - e^{-\tilde{t}/\tau}\right) + \beta_x^0.$$

Introducing this expression for β_x on the RHS of the equation of motion of the velocity component β_z leads to

$$\frac{d\beta_z}{d\tilde{t}} + \frac{\beta_z}{\tau} = \mathcal{E}\beta_x\cos(\tilde{t}) = \mathcal{E}\left[\mathcal{E}\frac{\tau}{1+\tau^2}\left(\cos(\tilde{t}) + \tau\sin(\tilde{t}) - e^{-\tilde{t}/\tau}\right) + \beta_x^0\right]\cos(\tilde{t}).$$

Only the term $\propto \mathcal{E}^2\cos^2(\tilde{t})$ on the RHS has a nonvanishing time average. With $\langle\cos^2\rangle = 1/2$ and for times $\tilde{t} \gg \tau$ we obtain the steady-state drift velocity $\langle\beta_z\rangle$ with

$$\langle\beta_z\rangle = \frac{\mathcal{E}^2}{2}\frac{\tau^2}{1+\tau^2}.$$

Note that this result neither depends on the initial condition $\beta_x(0)$ nor on $\beta_z(0)$. If, for example, the damping time is really as short as 5 fs and we have $\hbar\omega_0 = 1.5\,eV$, $\tau = \omega_0\,5\,\text{fs} = 11.2$ and the factor $\tau^2/(1+\tau^2) = 0.99$, which is only slightly less than 1 for the limit $\tau \to \infty$.

4.5 We have seen that the fundamental emission frequency $\tilde{\omega}_0$ for initial condition $\zeta_0 = 0$ follows the simple form of (4.63). Thus, our spectrometer or filter frequency of $0.985 \times 2\omega_0 = 2\tilde{\omega}_0$ immediately gives the condition for the angle θ included by the wavevector of light and the detection direction

$$\cos\theta = \left(1 - \frac{1}{0.985}\right)\frac{4}{\mathcal{E}^2} + 1,$$

which leads to $\theta = 20$ degrees for $\mathcal{E}^2 = 1$. This means that you will only detect a SHG signal on a cone around the wavevector of light with opening angle θ (for illustration of this cone see Fig. 7.26). Other spectrometer frequencies and/or other values of $\mathcal{E}^2$ merely change the value of θ. Due to relativistic effects, the intensity on this cone is modulated with the azimuthal angle φ. For $\zeta_0 \neq 0$, one still gets a cone, however, its axis is no longer parallel to K (see (4.61) and (4.62)), but lies in the xz-plane.

4.6 Inserting $\tilde{E}_0/\tilde{B}_0 = c_0$ from Sect. 2.2 into (4.78) and introducing the cyclotron frequency ω_c according to (4.45) immediately leads to (4.79) for the Schwinger field $\tilde{E}_0$.

Problems of Chapter 5

5.1 The total potential $V(x)$ resulting from the binding potential $U(x)$ plus the contribution of the laser electric field is shown in Figs. 5.3 and 5.9. In contrast to the rectangular potential well (see Fig. 5.4), the barrier *height* is also significantly reduced for the Coulomb potential. Indeed, for a certain field, the barrier-suppression field, the peak of the barrier on the LHS is as low as $-E_b$. A simple and straightforward curve discussion shows that, for the hydrogen potential, this happens at a peak laser electric field given by

$$\tilde{E}_0 = \frac{E_b^2}{4\,e}\frac{4\pi\epsilon_0}{e^2} = 3.2 \times 10^{10}\,\text{V/m}\,.$$

This field is comparable to that for unity Keldysh parameter, i.e., $\tilde{E}_0 = 2.8\times10^{10}\,\text{V/m}$ from Example 5.1 for $\hbar\omega_0 = 1.5\,eV$. Thus, for increasing intensity of light one actually has a transition from multiphoton absorption directly to above-barrier ionization (rather than to tunneling) for the hydrogen atom under these conditions. This is, however, not true for all types of atoms. For hydrogen atoms and mid-infrared excitation with, e.g., $\hbar\omega_0 = 0.15\,eV$, one would have the scenario: multiphoton absorption $\to$ electrostatic tunneling $\to$ above-barrier ionization.

As not only the barrier width is reduced with increasing laser electric field but also the barrier *height*, the ionization rate is drastically influenced for $\hbar\omega_0 = 1.5\,eV$. Hence, (5.13) largely *under*estimates the actual rate for the hydrogen atom, which can be obtained from exact numerical solutions of the time-independent Schrödinger equation or by the approximative Ammosov–Delone–Krainov theory [149–153] (also see discussion in Ref. [135]). Essentially, the prefactor Γ_{ion}^0 in (5.13) is replaced by a function of the atomic quantum numbers and the ionization potential as well as by a power of the instantaneous laser electric field $E(t)$. The exponential dependence

$$\Gamma_{\text{ion}}(t) \propto e^{-\frac{E_{\text{exp}}}{|E(t)|}}$$

remains, however, with $E_{\text{exp}} \propto E_b$.

5.2 We consider a "+" or "−" circularly polarized laser electric field propagating along z (spatial dependence suppressed) according to

$$\boldsymbol{E}(t) = \tilde{E}(t) \begin{pmatrix} +\cos(\omega_0 t + \phi) \\ \pm\sin(\omega_0 t + \phi) \\ 0 \end{pmatrix}\,.$$

For constant envelope $\tilde{E}(t) = \tilde{E}_0$, we have $|\boldsymbol{E}(t)| = \tilde{E}_0 = \text{const.}$, thus *the ionization rate* according to (5.13) is constant in time and *does not oscillate with the carrier frequency of light* ω_0. After ionization of the atom, Newton's law tells us that the *electron never returns to the nucleus* at $\boldsymbol{r} = (0, 0, 0)^T$. Consequently, *no harmonics are generated along the lines of our reasoning for linear polarization.*

5.3 The somewhat arbitrary threshold in Fig. 5.5 lies around $|E(t)|/E_{\exp} = 0.05$, which is ten times lower than for the peak electric field $\tilde{E}_0$ in Fig. 5.12. If one reduces $\tilde{E}_0$ in the calculation (according to (5.20) and (5.13)) to $\tilde{E}_0/E_{\exp} = 0.05$ (all other parameters as in Fig. 5.12), one indeed gets nearly 100% modulation of $N_e(t \to \infty)$ versus ϕ. However, the absolute electron yield becomes ridiculously low, i.e., $N_e(\infty)/N_{\text{atom}}^0 = 3 \times 10^{-8}$ for $\phi = 0$. Also note that one leaves the electrostatic regime for too low values of $\tilde{E}_0$ (see footnote in Sect. 5.3).

5.4 For the parameters of Figs. 5.7 and 5.8, the dipole moment d for a transition from the ground state of the well to its first excited state is given by $d \approx 18\% \, eL = 0.11 \, e \, \text{nm}$ (see Problem 3.2). With the peak laser electric field of $\tilde{E}_0 = 3 \times 10^{10} \, \text{V/m}$ corresponding to $\gamma_K \approx 1$ and with $\hbar\omega_0 = 1.5 \, eV$, this translates into $\Omega_R/\omega_0 = 2.2$. At this point, the two-level system approximation fails because of the massive tunneling out of the well visible in Fig. 5.7, whereas tunneling is of minor importance in Fig. 5.8 where $\gamma_K \approx 2$ and $\Omega_R/\omega_0 = 1.1$. Thus, for the conditions of Fig. 5.7, the two-level system approximation fails if the peak Rabi frequency Ω_R exceeds the carrier frequency of light ω_0. The validity of the two-level system approach would obviously extend for larger well widths L as γ_K remains unchanged whereas Ω_R increases proportional to L. In any case, this reminds us that there are nonlinear optical contributions from internal excitations of the well that are not at all accounted for by our semiclassical treatment of high-harmonic generation.

5.5 Figure 3.4 refers to transitions from one discrete state #1 into another discrete state #2, while Fig. 5.12 describes transitions from a discrete state into a *continuum* of unbound states. Thus, the two-level system coherence in Fig. 3.4, i.e., the amplitude of the superposition state composed of the two eigenstates ψ_1 and ψ_2 (see Sect. 3.2), oscillates with a single frequency, namely the transition frequency Ω. In contrast to this, for the ionization of atoms, we have a broad distribution of transition frequencies, which interfere destructively and lead to a partial apparent loss of coherence (Fig. 5.12). Hence, e.g., Rabi flopping no longer occurs.

Problems of Chapter 7

7.1 For space inversion, i.e. $r \to -r$, the LHS of (7.27) transforms as $j_{\text{pd}} \to -j_{\text{pd}}$. As the wavevector $K \to -K$ and the electric field $E \to -E$, the RHS of (7.27) $\to -$RHS. Thus, the photon-drag tensor $\mathcal{T}$ can be nonzero in an inversion-symmetric material. This is consistent with our reasoning in Sect. 2.5, where we have argued with the magnetic field of the laser. A similar reasoning holds for the SHG polarization in (7.32).

7.2 In the perturbative regime and within the range of validity of the acceleration theorem [145], Newton's second law together with the Lorentz force for an isotropic semiconductor with (constant) effective electron mass m_e and electron charge $-e$

$$\ddot{r} = \dot{v} = -e \, \frac{1}{m_e} \, (E + v \times B)$$

turns into

$$\ddot{\boldsymbol{r}} = \dot{\boldsymbol{v}} = -e\,\mathcal{M}_{\mathrm{e}}^{-1}\,(\boldsymbol{E} + \boldsymbol{v} \times \boldsymbol{B})$$

for reduced symmetry. Here we have introduced the (velocity-independent) 3×3 *inverse effective-mass tensor* $\mathcal{M}_{\mathrm{e}}^{-1}$ with components [22, 145]

$$\left(\mathcal{M}_{\mathrm{e}}^{-1}\right)_{ij} = \frac{1}{\hbar}\frac{\partial v_i}{\partial k_j}(\boldsymbol{0}) = \frac{1}{\hbar^2}\frac{\partial^2 E_{\mathrm{e}}}{\partial k_i \partial k_j}(\boldsymbol{0})\,.$$

$\boldsymbol{k}$ is the electron wavevector. For example, for cubic symmetry, a possible energy dispersion relation – which is parabolic for any given direction of $\boldsymbol{k}$ – is

$$E_{\mathrm{e}}(\boldsymbol{k}) = \frac{\hbar^2 k^2}{2\,m}\left(1 + \mathcal{W}\,\frac{k_x^2 k_y^2 + k_y^2 k_z^2 + k_z^2 k_x^2}{k^4}\right),$$

where the mass m and the dimensionless warping $\mathcal{W}$ are material parameters (see Fig. 7.27). Such a form can indeed be derived for the valence band from $\boldsymbol{k} \cdot \boldsymbol{p}$ perturbation theory [22] in the limit of small $|\boldsymbol{k}|$.

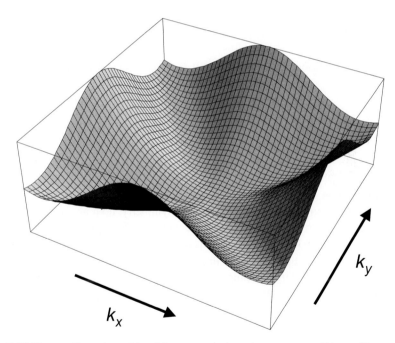

Fig. 7.27. Energy dispersion with cubic symmetry (warping parameter $\mathcal{W} = -3$) versus k_x and k_y for $k_z = 0$. Such a "warped" dispersion relation is typical for the valence band of many semiconductors. According to our reasoning in Sect. 7.4, the photon-drag current would still be parallel to the wavevector of light if the latter is oriented along the principal cubic axes or along the diagonals, i.e., if $\varphi = 0, \pi/4, \pi/2, \ldots$

Again using linear polarization of light according to

$$E(t) = \tilde{E}\,\cos(\omega_0 t + \phi) = \begin{pmatrix} \tilde{E}_1 \\ \tilde{E}_2 \\ \tilde{E}_3 \end{pmatrix} \cos(\omega_0 t + \phi)\,,$$

we obtain to *first order* in the laser electric field (see Sect. 4.4.1)

$$\boldsymbol{v} = -\frac{e}{\omega_0}\,\mathcal{M}_{\mathrm{e}}^{-1}\,\tilde{E}\,\sin(\omega_0 t + \phi)\,.$$

Note that, for reduced symmetry, $\boldsymbol{v}$ and E are generally no longer parallel to each other. To *second order* in the laser electric field, taking advantage of the fact that, for a plane wave according to the second Maxwell equation, the vectors E, B, and K follow

$$B(t) = \frac{1}{\omega_0}\,K \times E(t) = \frac{1}{c}\,\hat{K} \times E(t)\,,$$

with the unit wavevector of light $\hat{K} = K/|K|$ and the medium dispersion relation of light $|K| = \omega_0/c$, we have to solve the equation of motion

$$\ddot{r} = \dot{v} = -e\,\mathcal{M}_{\mathrm{e}}^{-1}\left[\left(-\frac{e}{\omega_0}\mathcal{M}_{\mathrm{e}}^{-1}\tilde{E}\right) \times \left(\frac{1}{c}\hat{K} \times \tilde{E}\right)\right]\underbrace{\sin(\omega_0 t + \phi)\cos(\omega_0 t + \phi)}_{= \frac{1}{2}\sin(2\omega_0 t + 2\phi)}\,.$$

For the initial conditions $\boldsymbol{v}(0) = 0$ and $r(0) = 0$ (see Sect. 4.4.1), we obtain

$$\boldsymbol{v}(t) = +\frac{e^2}{4\omega_0^2 c}\,\mathcal{M}_{\mathrm{e}}^{-1}\left[\left(\mathcal{M}_{\mathrm{e}}^{-1}\tilde{E}\right) \times \left(\hat{K} \times \tilde{E}\right)\right]\left(1 - \cos(2\omega_0 t + 2\phi)\right)$$

and

$$r(t) = +\frac{e^2}{4\omega_0^2 c}\,\mathcal{M}_{\mathrm{e}}^{-1}\left[\left(\mathcal{M}_{\mathrm{e}}^{-1}\tilde{E}\right) \times \left(\hat{K} \times \tilde{E}\right)\right]\left(t - \frac{1}{2\omega_0}\sin(2\omega_0 t + 2\phi)\right)\,.$$

Again, the motion consists of a constant drift, the photon-drag effect, and second-harmonic generation (SHG). But: For reduced symmetry, both are no longer necessarily parallel to the wavevector of light K. Details depend on the precise form of the inverse effective-mass tensor $\mathcal{M}_{\mathrm{e}}^{-1}$. It is clear, however, that the components of both, the photon-drag current as well as of the SHG are generally linear forms in $\hat{K}$ and quadratic forms in $\tilde{E}$. Thus, lumping all the prefactors into the coefficients T_{ijkl}, we can express the components of the photon-drag current-density vector according to

$$j_i = -\frac{eN_{\mathrm{e}}}{V}\,\langle v_i\rangle = \sum_{j,k,l=1}^{3} T_{ijkl}\,\hat{K}_j\,\tilde{E}_k\,\tilde{E}_l\,,$$

which is identical to (7.27). In analogy, the contribution of the optical polarization $P_i = -eN_{\mathrm{e}}/V\,r_i$ corresponding to SHG leads to (7.32). The reasoning for the hole contributions is analogous. It is the anisotropy of the hole dispersion that is actually relevant in the experiments on germanium [224].

Symbols

$a,$	lattice constant of a crystal
$a_e^0,$	peak electron acceleration
$\alpha,$	absorption coefficient
$\boldsymbol{A}(\boldsymbol{r}, t),$	vector potential
$\tilde{A}_0,$	peak of the vector potential envelope
$\boldsymbol{B}(\boldsymbol{r}, t),$	$\boldsymbol{B}$–field or magnetic field
$\tilde{B}_0,$	peak of magnetic field envelope
$\boldsymbol{\beta} = (\beta_x, \beta_y, \beta_z)^{\mathrm{T}} = \boldsymbol{v}/c_0,$	relativistic velocity
$c,$	medium speed of light
$c_0,$	vacuum speed of light with $c_0 = 2.9979 \times 10^8$ m/s
$\chi = \chi^{(1)},$	linear optical susceptibility
$\chi^{(N \neq 1)},$	nonlinear optical susceptibilities
$d,$	dipole matrix element
$d_{cv},$	dipole matrix element for valence band to conduction band transitions
$\boldsymbol{D}(\boldsymbol{r}, t),$	$\boldsymbol{D}$–field
$\mathcal{D},$	Hooke's spring constant
$\delta\omega\,\delta t,$	duration–bandwidth product, e.g., $\delta\omega\,\delta t = 2\pi \times 0.8859$ for $\mathrm{sinc}^2(t)$-pulses
$\Delta K,$	wavevector mismatch in harmonic generation
$\Delta\omega,$	mode spacing in a mode-locked laser oscillator
$\Delta\phi,$	pulse-to-pulse phase slip of the light field
$e,$	elementary charge with $e = +1.6021 \times 10^{-19}$ A s
$\mathrm{e},$	Euler's number with $\mathrm{e} = 2.7183$
$\boldsymbol{E}(\boldsymbol{r}, t),$	electric field
$\tilde{E}(t),$	electric-field envelope
$\tilde{E}_0,$	peak of electric-field envelope

E_b,	binding energy (or ionization potential)
E_e,	electron energy
E_g,	semiconductor bandgap energy
$\langle E_{kin} \rangle$,	ponderomotive energy (note that $\langle ... \rangle$ in this book can mean either the classical cycle average or the quantum-mechanical expectation value, depending on the context)
ϵ,	material dielectric function
ϵ_b,	background dielectric constant
ϵ_0,	vacuum dielectric constant with $\epsilon_0 = 8.8542 \times 10^{-12}\, \mathrm{A\,s\,V^{-1}m^{-1}}$
$\mathcal{E}$,	normalized electric-field strength (relativistic regime)
f,	frequency
f_1, f_2,	occupation numbers of the two-levels of a two-level system
f_e, f_h,	electron and hole occupation numbers, respectively
$f_\phi = 1/t_\phi$,	carrier-envelope offset (CEO) frequency
$f_r = 1/t_r$,	repetition frequency
$\boldsymbol{F}$,	force vector
ϕ,	carrier-envelope offset (CEO) phase
$\phi(\boldsymbol{r}, t)$,	electrostatic potential
φ_G,	Gouy phase
g,	gravitational acceleration near the earth's surface with $g = 9.81\, \mathrm{m/s^2}$
g,	gain coefficient
γ,	relativistic factor
γ_K,	Keldysh parameter
Γ_{ion},	ionization rate
$\hbar$,	Planck's constant $h/(2\pi)$, with $\hbar = 0.6582\, e\mathrm{V\,fs}$ or $\hbar = 1.0545 \times 10^{-34}\, \mathrm{J\,s}$
$\boldsymbol{H}(\boldsymbol{r}, t)$,	$\boldsymbol{H}$–field
$\mathcal{H}$,	Hamiltonian
i,	imaginary unit
I,	light intensity
I_{rad},	radiated light intensity
I_0,	reference intensity
$\boldsymbol{j}$,	electrical current density
j_{pd},	photon-drag current density
$\boldsymbol{k} = (k_x, k_y, k_z)^{\mathrm{T}}$,	electron wavevector
k_B,	Boltzmann's constant with $k_B = 1.3804 \times 10^{-23}\, \mathrm{J/K}$
$\boldsymbol{K} = (K_x, K_y, K_z)^{\mathrm{T}}$,	wavevector of light

l,	length (tunneling barrier width or sample thickness)
l_{coh},	coherence length
L,	length of the laser resonator
λ,	wavelength of light
λ_c,	electron Compton wavelength with $\lambda_c = 2.4262 \times 10^{-12}$ m
λ_e,	electron de Broglie wavelength
Λ,	modulation period in quasi phase-matching
m_0,	free electron (rest) mass with $m_0 = 9.1091 \times 10^{-31}$ kg
m_e,	effective crystal-electron mass or relativistic vacuum electron mass, respectively
m_h,	effective hole mass
M,	integer
$M(r, t)$,	magnetization
μ_0,	vacuum permeability with $\mu_0 = 4\pi \times 10^{-7}$ V s A^{-1} m^{-1}
n,	refractive index
n_2,	nonlinear refractive index
n_{eh},	electron hole density
N,	integer
N_e,	number of electrons
$\mathcal{O}$,	arbitrary quantum mechanical observable
ω,	spectrometer frequency
$\tilde{\omega}$,	normalized spectrometer frequency
ω_0,	carrier frequency of laser pulse
$\tilde{\omega}_0$,	fundamental emission frequency in nonlinear Thomson scattering
ω_0^e,	electron oscillation frequency in the relativistic regime
ω_c,	cyclotron frequency
ω_e,	electron frequency
ω_{pl},	plasma frequency
Ω,	optical (interband) transition frequency
Ω_B,	peak Bloch frequency
$\Omega_B(t)$,	instantaneous Bloch frequency
Ω_R,	peak Rabi frequency
$\Omega_R(t)$,	instantaneous Rabi frequency
$\tilde{\Omega}_R(t)$,	instantaneous envelope Rabi frequency
Ω_{tun},	peak tunneling "frequency"
p,	electron momentum
p_{vc},	transition amplitude for valence band to conduction band transitions
$P(r, t)$,	optical polarization

$\boldsymbol{r} = (x, y, z)^{\mathrm{T}}$,   coordinate vector
r_{B},   Bohr radius
R,   radius of curvature of Gaussian spherical beam profile
ρ,   electrical charge density

$\boldsymbol{S}(\boldsymbol{r}, t)$,   Poynting vector
S_{RF},   spectral radio-frequency power density
σ,   spectral "width" of a Gaussian

t,   time
$t_{\mathrm{FWHM}} = \delta t$,   full width at half-maximum of the temporal pulse intensity profile
t_{tun},   tunneling time
τ,   time delay of a Michelson interferometer
T,   temperature
T_1,   longitudinal relaxation time
T_2,   transverse relaxation time
θ,   angle included by the wavevector of light and the detection direction
Θ,   pulse area
$\tilde{\Theta}$,   envelope pulse area

$(u, v, w)^{\mathrm{T}}$,   Bloch vector
$U(t)$,   photomultiplier voltage
$U(x)$,   binding potential (energy)

$\boldsymbol{v}$,   velocity vector
V,   volume
$V(x)$,   potential energy

w,   inversion
$w(z)$,   "width" of transverse Gaussian profile
w_0,   beam waist

x,   classical displacement
x_0,   peak classical displacement
ξ,   dimensionless parameter proportional to the laser electric field

z,   coordinate
z_{f},   focal length
z_{R},   Rayleigh length

References

1. C. H. Townes: *How the Laser Happened – Adventures of a Scientist* (Oxford University Press, 1999)
2. N. Taylor: *LASER: The Inventor, the Nobel Laureate, and the Thirty-Year Patent War* (Kensington Publishing Corp., New York 2000)
3. T. M. Maiman: Nature **187**, 493 (1960)
4. P. A. Franken, A. E. Hill, C. W. Peters, and G. Weinreich: Phys. Rev. Lett. **7**, 118 (1961)
5. Y. R. Shen: *The Principles of Nonlinear Optics* (John Wiley & Sons 1984)
6. A. Yariv: *Quantum Electronics*, 3rd edition (John Wiley & Sons 1989)
7. L. Allen and J. H. Eberly: *Optical Resonance and Two-Level Atoms* (Dover Publications, New York 1987)
8. P. Meystre and M. Sargent III: *Elements of Quantum Optics*, 2nd edition (Springer, Berlin Heidelberg New York 1991)
9. R. W. Boyd: *Nonlinear Optics* (Academic Press, Boston 1992)
10. M. O. Scully and M. S. Zubairy: *Quantum Optics* (Cambridge University Press 1997)
11. L. E. Hargrove, R. L. Fork, and M. A. Pollack: Appl. Phys. Lett. **5**, 4 (1964)
12. D. E. Spence, P. N. Kean, and W. Sibbett: Opt. Lett. **16**, 42 (1991)
13. R. L. Fork, C. H. Brito Cruz, P. C. Becker, and C. V. Shank: Opt. Lett. **12**, 483 (1986)
14. U. Keller: Appl. Phys. B: Lasers Opt. **58**, 347 (1994)
15. P. M. W. French: Contemp. Phys. **37**, 283 (1996)
16. D. Strickland and G. Mourou: Opt. Commun. **56**, 219 (1985)
17. M. A. Perry and G. Mourou: Science **264**, 917 (1994)
18. S. Bahk, P. Rousseau, T. Planchon, V. Chvykov, G. Kalintchenko, A. Maksimchuk, G. Mourou, and V. Yanovsky: International Conference on Lasers and Electro Optics (CLEO), San Francisco (USA), May 16–21 2004, postdeadline paper CPDA5, conference digest
19. P. M. Dirac: Nature **192**, 235 (1937)
20. A. Y. Potekhin, A. V. Ivanchik, D. A. Varshalovich, K. M. Lanzetta, J. A. Baldwin, G. M. Williger, and R. F. Carswell: Astrophys. J. **505**, 523 (1998)
21. J. K. Webb, M. T. Murphy, V. V. Flambaum, V. A. Dzuba, J. D. Barrow, C. W. Churchill, J. X. Prochaska, and A. M. Wolfe: Phys. Rev. Lett. **87**, 091301 (2001)
22. W. Schäfer and M. Wegener: *Semiconductor Optics And Transport Phenomena*, Advanced Texts in Physics (Springer, Berlin Heidelberg New York 2002)
23. K. L. Sala, G. A. Kenney-Wallace, and G. E. Hall: IEEE J. Quantum Electron. **16**, 990 (1980)

24. E. V. Baklanov, and V. P. Chebotaev: Sov. J. Quantum Electron. **7**, 1252 (1977)
25. H. R. Telle, G. Steinmeyer, A. E. Dunlop, J. Stenger, D. H. Sutter, and U. Keller: Appl. Phys. B **69**, 327 (1999)
26. S. T. Cundiff, J. Ye, and J. L. Hall: Rev. Sci. Instrum. **72**, 3749 (2001)
27. Th. Udem, R. Holzwarth, and T. W. Hänsch: Nature **416**, 233 (2001)
28. S. T. Cundiff and J. Ye: Rev. Mod. Phys. **75**, 325 (2003)
29. S. Karshenboim: Can. J. Phys. **78**, 639 (2000)
30. Th. Udem, S. A. Diddams, K. R. Vogel, C. W. Oates, E. A. Curtis, W. D. Lee, W. M. Itano, R. E. Drullinger, J. C. Bergquist, and L. Hollberg: Phys. Rev. Lett. **86**, 4996 (2001)
31. Th. Damour, F. Piazza, and G. Veneziano: Phys. Rev. Lett. **89**, 081601 (2002)
32. R. J. Rafac, B. C. Young, J. A. Beall, W. M. Itano, D. J. Wineland, and J. C. Bergquist: Phys. Rev. Lett. **85**, 2462 (2002)
33. L. B. Madsen: Phys. Rev. A **65**, 053417 (2002)
34. M. Fiebig, D. Fröhlich, B. B. Krichevtsov, and R. V. Pisarev: Phys. Rev. Lett. **73**, 2127 (1994)
35. A. Apolonski, A. Poppe, G. Tempea, Ch. Spielmann, Th. Udem, R. Holzwarth, T. W. Hänsch, and F. Krausz: Phys. Rev. Lett. **85**, 740 (2000)
36. D. J. Jones, S. A. Diddams, J. K. Ranka, A. Stentz, R. S. Windeler, J. L. Hall, and S. T. Cundiff: Science **288**, 635 (2000)
37. U. Morgner, R. Ell, G. Metzler, T. R. Schibli, F. X. Kärtner, J. G. Fujimoto, H. A. Haus, and E. P. Ippen: Phys. Rev. Lett. **86**, 5462 (2001)
38. A. Baltuska, T. Fuji, and T. Kobayashi: Phys. Rev. Lett. **88**, 133901 (2002)
39. T. Fuji, A. Apolonski, and F. Krausz: Opt. Lett. **29**, 632 (2004)
40. C. W. Luo, K. Reimann, M. Woerner, T. Elsaesser, R. Hey, and K. H. Ploog: Phys. Rev. Lett. **92**, 047402 (2004)
41. C. G. B. Garrett and D. E. Mc Cumber: Phys. Rev. A **1**, 305 (1970)
42. A. Puri and J. L. Birman: Phys. Rev. A **27**, 1044 (1983)
43. S. Chu and S. Wong: Phys. Rev. Lett. **48**, 738 (1982)
44. J. Peatross, S. A. Glasgow, and M. Ware: Phys. Rev. Lett. **84**, 2370 (2000)
45. L. Brillouin: *Wave Propagation and Group Velocity* (Academic Press, New York 1960)
46. K. E. Oughstun and C. M. Balictsis: Phys. Rev. Lett. **77**, 2210 (1996)
47. G. S. Sherman and K. E. Oughstun: Phys. Rev. A **41**, 6090 (1990)
48. K. E. Oughstun and G. S. Sherman: J. Opt. Soc. Am. B **5**, 817 (1988)
49. P. Pleshko and I. Palocz: Phys. Rev. Lett. **22**, 1201 (1969)
50. T. Ankel: Z. Phys. B **144**, 120 (1956)
51. F. Bloch: Phys. Rev. **70**, 460 (1946)
52. F. Bloch, W. W. Hansen, and M. Packard: Phys. Rev. **70**, 960 (1946)
53. I. I. Rabi: Phys. Rev. **51**, 652 (1937)
54. O. D. Mücke: *Extreme Nonlinear Optics in Semiconductors with Intense Two-Cycle Laser Pulses*, PhD thesis, Universität Karlsruhe (TH) (Shaker Verlag 2003)
55. F. Bloch and A. Siegert: Phys. Rev. **57**, 522 (1940)
56. B. R. Mollow: Phys. Rev. **188**, 1969 (1969)
57. B. R. Mollow: Phys. Rev. A **2**, 76 (1970)
58. B. R. Mollow: Phys. Rev. A **5**, 2217 (1972)
59. F. Y. Wu, S. Ezekiel, M. Ducloy, and B. R. Mollow: Phys. Rev. Lett. **38**, 1077 (1977)
60. S. Hughes: Phys. Rev. Lett. **81**, 3363 (1998)
61. S. Hughes: Phys. Rev. A **62**, 055401 (2000)
62. R. Bavli and H. Metiu: Phys. Rev. Lett. **69**, 1986 (1992)
63. M. Yu. Ivanov, P. B. Corkum, and P. Dietrich: Laser Phys. **3**, 375 (1993)

64. A. Levinson, M. Segev, G. Almogy, and A. Yariv: Phys. Rev. B **49**, R 661 (1994)
65. T. Zuo, S. Chelkowski, and A. D. Bandrauk: Phys. Rev. A **49**, 3943 (1994)
66. R. W. Ziolkowski, J. M. Arnold, and D. M. Gogny: Phys. Rev. A **52**, 3082 (1995)
67. T. Tritschler, O. D. Mücke, M. Wegener, U. Morgner, and F. X. Kärtner: Phys. Rev. Lett. **90**, 217404 (2003)
68. E. Dupont, P. B. Corkum, H. C. Liu, M. Buchanan, and Z. R. Wasilewski: Phys. Rev. Lett. **74**, 3596 (1995)
69. R. Atanosov, A. Hache, J. L. P. Hughes, H. M. van Driel, and J. E. Sipe: Phys. Rev. Lett. **76**, 1703 (1996)
70. A. Hache, Y. Kostoulas, R. Atanosov, J. L. P. Hughes, J. E. Sipe, and H. M. van Driel: Phys. Rev. Lett. **78**, 306 (1997)
71. R. D. R. Bhat and J. E. Sipe: Phys. Rev. Lett. **85**, 5432 (2000)
72. M. J. Stevens, A. L. Smirl, R. D. R. Bhat, A. Najmaie, J. E. Sipe, and H. M. van Driel: Phys. Rev. Lett. **90**, 136603 (2003)
73. J. Hübner, W. W. Rühle, M. Klude, D. Hommel, R. D. R. Bhat, J. E. Sipe, and H. M. van Driel: Phys. Rev. Lett. **90**, 216601 (2003)
74. T. M. Fortier, P. A. Roos, D. J. Jones, S. T. Cundiff, R. D. R. Bhat, and J. E. Sipe: Phys. Rev. Lett. **92**, 147403 (2004)
75. J. V. Moloney and W. J. Meath: Phys. Rev A **17**, 1550 (1978)
76. G. F. Thomas: Phys. Rev. A **32**, 1515 (1985)
77. W. M. Griffith, M. W. Noel, and T. F. Gallagher: Phys. Rev. A **57**, 3698 (1998)
78. A. Brown and W. J. Meath: J. Chem. Phys. **109**, 9351 (1998)
79. T. Tritschler, O. D. Mücke and M. Wegener: Phys. Rev. A **68**, 033404 (2003)
80. M. Abramowitz and I. A. Stegun: *Handbook of Mathematical Functions*, 9th printing (Dover Publications Inc., New York 1979)
81. J. H. Shirley: Phys. Rev. **138**, B 979 (1965)
82. T. Hattori and T. Kobayashi: Phys. Rev. A **35**, 2733 (1987)
83. S. M. Barnett, P. Filipowicz, J. Javanainen, P. L. Knight, and P. Meystre: p. 485, in *Frontiers in Quantum Optics*, E. R. Pike and S. Sarkar, eds. (Adam Hilger, Bristol 1986)
84. G. M. Genkin: Phys. Rev. A **58**, 758 (1998)
85. R. Parzynski and M. Sobczak: Opt. Commun. **228**, 111 (2003)
86. A. P. Jauho and K. Johnsen: Phys. Rev. Lett. **76**, 4576 (1996)
87. K. Johnsen and A. P. Jauho: Phys. Rev. B **57**, 8860 (1998)
88. A. H. Chin, J. M. Bakker, and J. Kono: Phys. Rev. Lett. **85**, 3293 (2000)
89. H. R. Reiss: Phys. Rev. A **19**, 1140 (1979)
90. H. R. Reiss: Phys. Rev. A **22**, 1786 (1980)
91. D. M. Volkov: Z. Physik **94**, 250 (1935)
92. H. D. Jones and H. R. Reiss: Phys. Rev. B **16**, 2466 (1977)
93. A. V. Jones and G. J. Papadopoulos: J. Phys. A **4**, L87 (1971)
94. *Landolt-Börnstein*, Vols. III/17a and III/17b (Springer-Verlag)
95. F. Bloch: Z. Phys. **52**, 555 (1928)
96. C. Zener: Proc. R. Soc. London Ser. A **145**, 523 (1934)
97. G. H. Wannier: Phys. Rev. **100**, 1227 (1955)
98. G. H. Wannier: Phys. Rev. **117**, 432 (1960)
99. G. H. Wannier: Rev. Mod. Phys. **34**, 645 (1962)
100. L. Esaki and R. Tsu: Appl. Phys. Lett. **19**, 246 (1971)
101. A. A. Ignatov and Y. A. Romanov: Sov. Phys. Solid State **17**, 2216 (1975)
102. A. A. Ignatov and Y. A. Romanov: Phys. Stat. Sol. B **73**, 327 (1976)
103. M. W. Feise and D. S. Citrin: Appl. Phys. Lett. **75**, 3536 (1999)

212   References

104. S. Winnerl, E. Schomburg, S. Brandl, O. Kus, K. F. Renk, M. C. Wanke, S. J. Allen, A. A. Ignatov, V. Ustinov, A. Zhukov, and P. S. Kop'ev: Appl. Phys. Lett. **77**, 1259 (2000)
105. F. V. Vachaspati: Phys. Rev. **128**, 664 (1962)
106. L. S. Brown and T. W. B. Kibble: Phys. Rev. **133**, A 705 (1964)
107. E. S. Sarachik and G. T. Schappert: Phys. Rev. D **1**, 2738 (1970)
108. J. E. Gunn and J. P. Ostriker: Astrophys. J. **165**, 523 (1971)
109. E. Esarey, S. K. Ride and P. Sprangle: Phys. Rev. E **48**, 3003 (1993)
110. S.-Y. Chen, A. Maksimchuk, and D. Umstadter: Nature **396**, 653 (1998)
111. F. He, Y. Lau, D. P. Umstadter, and T. Strickler: Phys. Plasmas **9**, 4325 (2002)
112. F. He, Y. Y. Lau, D. P. Umstadter, and R. Kowalczyk: Phys. Rev. Lett. **90**, 055002 (2003)
113. Y. Y. Lau, F. He, D. P. Umstadter, and R. Kowalczyk: Phys. Plasmas **10**, 2155 (2003)
114. J. D. Jackson: *Classical Electrodynamics*, p. 480 (Wiley, New York 1962)
115. H. K. Avetissian, A. K. Avetissian, G. F. Mkrtchian, and Kh. V. Sedrakian: Phys. Rev. E **66**, 016502 (2002)
116. D. L. Burke, R. C. Field, G. Horton-Smith, J. E. Spencer, D. Walz, S. C. Berridge, W. M. Bugg, K. Shmakov, A. W. Weidemann, C. Bula, K. T. Mc Donald, E. J. Prebys, C. Bamber, S. J. Boege, T. Koffas, T. Kotseroglou, A. C. Melissonos, D. D. Myerhofer, D. A. Reis, and W. Ragg: Phys. Rev. Lett. **79**, 1626 (1997)
117. J. Schwinger: Phys. Rev. **82**, 664 (1951)
118. J. Schwinger: Phys. Rev. **93**, 615 (1954)
119. E. Brezin and C. Itzykon: Phys. Rev. D **2**, 1191 (1970)
120. T. Tajima and G. Mourou: Phys. Rev. Special Topics **5**, 031301 (2002)
121. S. W. Hawking: Nature **248**, 30 (1974)
122. W. G. Unruh: Phys. Rev. D **14**, 870 (1976)
123. E. Yablonovitch: Phys. Rev. Lett. **62**, 1742 (1989)
124. P. Chen and T. Tajima: Phys. Rev. Lett. **83**, 256 (1999)
125. G. Farkas and C. Tóth: Phys. Lett. A **168**, 447 (1992)
126. S. E. Harris, J. J. Macklin, and T. W. Hänsch: Opt. Commun. **100**, 487 (1993)
127. L. V. Keldysh: Sov. Phys. JETP **20**, 1307 (1965)
128. G. G. Paulus, F. Grasbon, H. Walther, P. Villoresi, M. Nisoli, S. Stagira, E. Priori, and S. De Silvestri: Nature **414**, 182 (2001)
129. J. Gao, F. Shen, and J. G. Eden: Phys. Rev. Lett. **81**, 1833 (1998)
130. J. Gao, F. Shen, and J. G. Eden: Phys. Rev. A **61**, 043812 (2000)
131. J. Gao, F. Shen, and J. G. Eden: Int. J. Mod. Phys. B **14**, 889 (2000)
132. K. Boyer, T. S. Luk, and C. K. Rhodes: Phys. Rev. Lett. **60**, 557 (1988)
133. K. W. D. Ledingham, I. Spencer, T. McCanny, R. P. Singhal, M. I. K. Santala, E. Clark, I. Watts, F. N. Beng, M. Zepf, K. Krushelnick, M. Tatarakis, A. E. Dangor, P. A. Norreys, R. Allott, D. Neely, R. J. Clark, A. C. Machacek, J. S. Wark, A. J. Cresswell, D. C. W. Sanderson, and J. Magill: Phys. Rev. Lett. **84**, 899 (2000)
134. T. E. Cowan, A. W. Hunt, T. W. Phillips, S. C. Wilks, M. D. Perry, C. Brown, W. Fountain, S. Hatchett, J. Johnson, M. H. Key, T. Parnell, D. M. Pennington, R. A. Snavely, and Y. Takahashi: Phys. Rev. Lett. **84**, 903 (2000)
135. T. Brabec and F. Krausz: Rev. Mod. Phys. **72**, 545 (2000)
136. W. Becker, S. Long, and J. M. MicIver: Phys. Rev. A **R 41**, 4112 (1990)
137. J. L. Krause, K. L. Schafer, and K. C. Kulander: Phys. Rev. Lett. **68**, 3535 (1992)
138. A. Pukhov, S. Gordienko, and T. Baeva: Phys. Rev. Lett. **91**, 173002 (2003)
139. T. F. Gallagher: Phys. Rev. Lett. **61**, 2304 (1988)
140. P. B. Corkum, N. H. Burnett, and F. Brunel: Phys. Rev. Lett. **62**, 1259 (1989)
141. P. B. Corkum: Phys. Rev. Lett. **71**, 1994 (1993)

142. M. Lewenstein, Ph. Balcou, M. Yu. Ivanov, A. L'Hullier, and P. B. Corkum: Phys. Rev. A **49**, 2117 (1994)
143. L. C. Dinu, H. G. Muller, S. Kazamias, G. Mullot, F. Auge, Ph. Balcou, P. M. Paul, M. Kovacev, P. Breger, and P. Agostini: Phys. Rev. Lett. **91**, 063901 (2003)
144. C. H. Keitel and S. X. Hu: Appl. Phys. Lett. **80**, 541 (2003)
145. N. W. Ashcroft and N. D. Mermin: *Solid State Physics* (Saunders College Publishing 1976)
146. C. Lemell, X.-M. Tong, F. Krausz, and J. Burgdörfer: Phys. Rev. Lett. **90**, 076403 (2003)
147. A. Apolonski, P. Dombi, G. G. Paulus, M. Kakehata, R. Holzwarth, Th. Udem, Ch. Lemell, K. Torizuka, J. Burgdörfer, T. W. Hänsch, and F. Krausz: Phys. Rev. Lett. **92**, 073902 (2004)
148. P. Dombi, A. Apolonski, Ch. Lemell, G. G. Paulus, M. Kakehata, R. Holzwarth, Th. Udem, K. Torizuka, J. Burgdörfer, T. W. Hänsch, and F. Krausz: New J. Phys. **6**, 39 (2004)
149. M. V. Ammosov, N. B. Delone, and V. P. Krainov: Sov. Phys. JETP **64**, 1191 (1986)
150. V. P. Krainov: J. Opt. Soc. Am. **B 14**, 425 (1997)
151. B. Walker, B. Sheehy, L. F. DiMauro, P. Agostini, K. J. Schafer, and K. C. Kulander: Phys. Rev. Lett. **73**, 1227 (1994)
152. E. A. Chowdhury, C. P. J. Barty, and B. C. Walker: Phys. Rev. A **63**, 042712 (2001)
153. K. Yamakawa, Y. Akahane, Y. Fukuda, M. Aoyama, N. Inoue, H. Ueda, and T. Utsumi: Phys. Rev. Lett. **92**, 123001 (2004)
154. T. Brabec and F. Krausz: Phys. Rev. Lett. **78**, 3282 (1997)
155. K. L. Shlager and J. B. Schneider: IEEE Antennas Propagat. Mag. **37**, 39 (1995)
156. K. S. Yee: IEEE Trans. Antennas Propagat. **14**, 302 (1966)
157. J. P. Berenger: J. Comp. Phys. **114**, 185 (1994)
158. E. L. Lindman: J. Comp. Phys. **18**, 66 (1975)
159. A. E. Siegmann: *Lasers* (University Science, Mill Valley, Calif. 1986)
160. C. R. Gouy: Acad. Sci. Paris **110**, 1251 (1890)
161. C. R. Gouy: Ann. Chim. Phys. Ser. 6, **24**, 145 (1891)
162. S. Feng and H. G. Winful: Opt. Lett. **26**, 485 (2001)
163. M. A. Porras: Phys. Rev. E **65**, 026606 (2002)
164. R. W. Ziolkowski: Phys. Rev. A **39**, 2005 (1989)
165. Z. Wang, Z. Zhang, Z. Xu, and Q. Lin: IEEE J. Quantum Electron. **33**, 566 (1997)
166. S. Feng, H. G. Winful, and R. W. Hellwarth: Phys. Rev. E **59**, 4630 (1999)
167. A. E. Kaplan: J. Opt. Soc. Am. B **15**, 951 (1998)
168. P. Saari: Opt. Exp. **8**, 590 (2001)
169. A. B. Ruffin, J. V. Rudd, J. F. Whitaker, S. Feng, and H. G. Winful: Phys. Rev. Lett. **83**, 3410 (1999)
170. E. Budiarto, N.-W. Pu, S. Jeong, and J. Bokor: Opt. Lett. **23**, 213 (1998)
171. F. Lindner, G. G. Paulus, H. Walther, A. Baltuska, E. Goulielmakis, M. Lezius, and F. Krausz: Phys. Rev. Lett. **92**, 113001 (2004)
172. Z. L. Horvath and Zs. Bor: Phys. Rev. E **60**, 2337 (1999)
173. D. Du, X. Liu, G. Korn, J. Squier, and G. Mourou: Appl. Phys. Lett. **64**, 3071 (1994)
174. B. C. Stuart, D. Feit, A. M. Rubenchik, B. W. Shore, and M. D. Perry: Phys. Rev. Lett. **74**, 2248 (1995)
175. M. Lenzner, J. Krüger, S. Sartania, Z. Cheng, Ch. Spielmann, G. Mourou, W. Kautek, and F. Krausz: Phys. Rev. Lett. **80**, 4076 (1998)
176. A. C. Tien, S. Backus, H. Kapteyn, and M. Murnane: Phys. Rev. Lett. **82**, 3883 (1999)
177. U. Morgner, F. X. Kärtner, S. H. Cho, Y. Chen, H. A. Haus, J. G. Fujimoto, E. P. Ippen, V. Scheuer, G. Angelow, and T. Tschudi: Opt. Lett. **24**, 411 (1999)

178. M. U. Wehner, M. H. Ulm, and M. Wegener: Opt. Lett. **22**, 1455 (1997)

179. The first microscope objective has a focal length of 5.41 mm and a numerical aperture of NA=0.5 (*Coherent 25-0522*), the second one 13.41 mm and NA=0.5 (*Coherent 25-0555*). In these experiments, one loses about a factor of two in average power on the first microscope objective. This is due to the fact that the beam diameter is chosen to be larger than the objective aperture in order to get to the minimum spot radius. Thus, all relevant powers are consistently quoted *in front of the sample*.

180. N. Peyghambarian, S. W. Koch, and A. Mysyrowicz: *Introduction to Semiconductor Optics* (Prentice Hall, Englewood Cliffs, New Jersey 1993)

181. O. D. Mücke, T. Tritschler, M. Wegener, U. Morgner, and F. X. Kärtner: Phys. Rev. Lett. **87**, 057401 (2001)

182. Q. T. Vu, L. Bányai, H. Haug, O. D. Mücke, T. Tritschler, and M. Wegener: Phys. Rev. Lett. **92**, 217403 (2004)

183. M. Sheik-Bahae, A. A. Said, T.-H. Wei, D. J. Hagan, and E. W. van Stryland: IEEE J. Quantum Electron. **26**, 760 (1990)

184. L. Bányai, D. B. Tran Thoai, E. Reitsamer, H. Haug, D. Steinbach, M. U. Wehner, M. Wegener, T. Marschner, and W. Stolz: Phys. Rev. Lett. **75**, 2188 (1995)

185. M. U. Wehner, M. H. Ulm, D. S. Chemla, and M. Wegener: Phys. Rev. Lett. **80**, 1992 (1998)

186. W. A. Hügel, M. F. Heinrich, and M. Wegener, Q. T. Vu, L. Bányai, and H. Haug: Phys. Rev. Lett. **83**, 3313 (1999)

187. Q. T. Vu, H. Haug, W. A. Hügel, S. Chatterjee, and M. Wegener: Phys. Rev. Lett. **85**, 3508 (2000)

188. H. Haug: Nature **414**, 261 (2001)

189. R. Huber, F. Tauser, A. Brodschelm, M. Bichler, G. Abstreiter, and A. Leitenstorfer: Nature **414**, 286 (2001)

190. W. A. Hügel, M. Wegener, Q. T. Vu, L. Bányai, H. Haug, F. Tinjod, and H. Mariette: Phys. Rev. B **66**, 153203 (2002)

191. K. Leo, M. Wegener, J. Shah, D. S. Chemla, E. O. Göbel, T. C. Damen, S. Schmitt-Rink, and W. Schäfer: Phys. Rev. Lett. **65**, 1340 (1990)

192. M. Wegener, D. S. Chemla, S. Schmitt-Rink, and W. Schäfer: Phys. Rev. A **42**, 5675 (1990)

193. R. Binder, S. W. Koch, M. Lindberg, N. Peyghambarian, and W. Schäfer: Phys. Rev. Lett. **65**, 899 (1990)

194. S. T. Cundiff, A. Knorr, J. Feldmann, S. W. Koch, E. O. Göbel, and H. Nickel: Phys. Rev. Lett. **73**, 1178 (1994)

195. H. Giessen, A. Knorr, S. Haas, S. W. Koch, S. Linden, J. Kuhl, M. Hetterich, M. Grün, and C. Klingshirn: Phys. Rev. Lett. **81**, 4260 (1998)

196. A. Schülzgen, R. Binder, M. E. Donovan, M. Lindberg, K. Wundke, H. M. Gibbs, G. Khitrova, and N. Peyghambarian: Phys. Rev. Lett. **82**, 2346 (1999)

197. L. Bányai, Q. T. Vu, B. Mieck, and H. Haug: Phys. Rev. Lett. **81**, 882 (1998)

198. C. Ciuti, C. Piermarocchi, V. Savona, P. E. Selbmann, P. Schwendimann, and A. Quattropani: Phys. Rev. Lett. **84**, 1752 (2000)

199. H. Haug and S. W. Koch: *Quantum Theory of the Optical and Electronic Properties of Semiconductors*, 2nd edition (World Scientific 1993)

200. The fit formula for the envelope of the electric field spectrum is given by the sum of three Gaussians, i.e., by $|\tilde{E}_{\omega 0}(\omega)| = \sum_{n=1}^{3} E_n \exp(-(\omega - \omega_n)^2/\sigma_n^2)$, with the parameters $E_2/E_1 = 0.72$, $E_3/E_1 = 1.16$ and E_1 being determined by $\tilde{E}_0$; $\hbar\omega_1 = 1.38\,eV$, $\hbar\omega_2 = 1.68\,eV$, and $\hbar\omega_3 = 1.82\,eV$; $\hbar\sigma_1 = 0.10\,eV$, $\hbar\sigma_2 = 0.19\,eV$, and $\hbar\sigma_3 = 0.03\,eV$. For

the ZnO calculations, in order to avoid artifacts, the low and high-energy Gaussian tails of this spectrum (that predominantly arise from the broad central Gaussian) are suppressed by an analytic function. The real electric field $E(t)$ of an individual pulse results from the real part of the Fourier transform of $|\tilde{E}_{\omega_0}(\omega)|$. Note that the CEO phase ϕ of $E(t)$ can be modified without explicitly decomposing it into carrier wave $\cos(\omega_0 t + \phi)$ and envelope $\tilde{E}(t)$ – which would not be possible analytically anyway. Also, see Problem 2.5.

201. D. E. Aspnes, S. M. Kelso, R. A. Logan, and R. Bhat: J. Appl. Phys. **60**, 754 (1986)
202. O. D. Mücke, T. Tritschler, M. Wegener, U. Morgner, and F. X. Kärtner: Phys. Rev. Lett. **89**, 127401 (2002)
203. V. F. Elesin: Sov. Phys. JETP **32**, 328 (1971)
204. C. Comte and G. Mahler: Phys. Rev. B **34**, 7164 (1986)
205. S. Schmitt-Rink, D. S. Chemla, and H. Haug: Phys. Rev. B **37**, 941 (1988)
206. F. Jahnke and K. Henneberger: Phys. Rev. B **45**, 4077 (1992)
207. V. Skrikand, and D. R. Clarke: J. Appl. Phys. **83**, 5447 (1998)
208. O. D. Mücke, T. Tritschler, M. Wegener, U. Morgner, and F. X. Kärtner: Opt. Lett. **27**, 2127 (2002)
209. X. W. Sun and H. S. Kwok: J. Appl. Phys. **86**, 408 (1999)
210. K. Postava, H. Sueki, M. Aoyama, T. Yamaguchi, Ch. Ino, Y. Igasaki, and M. Horie: J. Appl. Phys. **87**, 7820 (2000)
211. W. Franz: Z. Naturforschg. A **13**, 484 (1958)
212. L. V. Keldysh: Sov. Phys. JETP **34**, 788 (1958)
213. Y. Yacobi: Phys. Rev. **169**, 610 (1968)
214. A. Srivastava and J. Kono: International Conference on *Quantum Electronics and Laser Science* (QELS), Baltimore (USA), June 1–6 2003, paper QFD2, conference digest (2003)
215. A. H. Chin, O. G. Calderon, and J. Kono: Phys. Rev. Lett. **86**, 3292 (2001)
216. K. B. Nordstrom, K. Johnsen, S. J. Allen, A.-P. Jauho, B. Birnir, J. Kono, T. Noda, H. Akiyama, and H. Sakaki: Phys. Rev. Lett. **81**, 457 (1998)
217. J. P. Gordon: Phys. Rev. A **8**, 14 (1973)
218. R. Peierls: Proc. R. Soc. Lond. A **347**, 475 (1976)
219. R. Peierls: Proc. R. Soc. Lond. A **355**, 141 (1977)
220. H. M. Barlow: Nature **173**, 41 (1954)
221. W. Lehr and R. von Baltz: Z. Phys. B **51**, 25 (1983)
222. A. F. Gibson, M. F. Kimmitt, and A. C. Walker: Appl. Phys. Lett. **17**, 75 (1970)
223. A. M. Danishevskii, A. A. Kastal'skii, S. M. Ryvkin, and I. D. Yaroshetskii: Sov. Phys. JETP **31**, 292 (1979)
224. A. C. Walker and D. R. Tilley: J. Phys. C **4**, L376 (1971)
225. K. D. Moll, D. Homoelle, A. L. Gaeta, and R. W. Boyd: Phys. Rev. Lett. **88**, 153901 (2002)
226. M. Drescher, M. Hentschel, R. Kienberger, G. Tempea, C. Spielmann, G. A. Reider, P. B. Corkum, and F. Krausz: Science **291**, 1923 (2001)
227. M. Hentschel, R. Kienberger, Ch. Spielmann, G. A. Reider, N. Milosevic, T. Brabec, P. Corkum, U. Heinzmann, M. Drescher, and F. Krausz: Nature **414**, 509 (2001)
228. N. A. Papadogiannis, B. Witzel, C. Kalpouzos, and D. Charalambidis: Phys. Rev. Lett. **83**, 4289 (1999)
229. P. M. Paul, E. S. Toma, P. Breger, G. Mullot, F. Augé, Ph. Balcou, H. G. Muller, and P. Agostini: Science **292**, 1689 (2001)
230. A. Rundquist, C. G. Durfee III, Z. Chang, C. Herne, S. Backus, M. M. Murnane, and H. C. Kapteyn: Science **280**, 1412 (1998)
231. C. G. Durfee, A. R. Rundquist, S. Backus, C. Herne, M. M. Murnane, and H. C. Kapteyn: Phys. Rev. Lett. **83**, 2187 (1999)

232. A. Paul, R. A. Bartels, R. Tobey, H. Green, S. Weiman, I. P. Christov, M. M. Murnane, H. C. Kapteyn, and S. Backus: Nature **421**, 51 (2003)

233. A. McPherson, G. Gibson, H. Jara, U. Johann, T. S. Luk, I. A. McIntyre, K. Boyer, and C. K. Rhodes: J. Opt. Soc. Am. B **4**, 595 (1987)

234. M. Ferray, A. L'Hullier, X. F. Li, A. Lompre, G. Mainfray, and C. Manus: J. Phys. B **21**, L 31 (1988)

235. X. F. Li, A. L'Hullier, M. Ferray, L. A. Lompre, and G. Mainfray: Phys. Rev. A **39**, 5751 (1991)

236. N. Sarukura, K. Hata, T. Adachi, R. Nodomi, M. Watanabe, and S. Watanabe: Phys. Rev. A **43**, 1669 (1991)

237. Y. Akiyami, K. Midorikawa, Y. Matsunawa, Y. Nagata, M. Obara, H. Tashiro, and K. Toyoda: Phys. Rev. Lett. **69**, 2176 (1992)

238. J. J. Macklin, J. D. Kmetec, and C. L. Grodon III: Phys. Rev. Lett. **70**, 766 (1993)

239. A. L'Huillier and Ph. Balcou: Phys. Rev. Lett. **70**, 774 (1993)

240. K. Miyazaki, H. Sakai, G. U. Kim, and H. Takada: Phys. Rev. A **49**, 548 (1994)

241. J. G. W. Tisch, R. A. Smith, J. E. Muffett, M. Ciarocca, J. P. Marangos, and M. H. R. Hutchinson: Phys. Rev. A **49**, R 28 (1994)

242. K. Miyazaki and H. Takada: Phys. Rev. A **52**, 3007 (1995)

243. Z. Chang, A. Rundquist, H. Wang, M. M. Murnane, and H. C. Kapteyn: Phys. Rev. Lett. **79**, 2967 (1997)

244. Ch. Spielmann, N. H. Burnett, S. Sartania, R. Koppitsch, M. Schnürer, C. Kan, M. Lenzner, P. Wobrauschek, and F. Krausz: Science **278**, 661 (1997)

245. P. Salieres, A. L'Huillier, P. Antoine, and M. Lewenstein: Adv. At., Mol., Opt. Phys. **41**, 83 (1999)

246. C. J. Joachain, M. Dörr, and N. J. Kylstra: Adv. At., Mol., Opt. Phys. **42**, 225 (2000)

247. E. Seres, J. Seres, F. Krausz, and C. Spielmann: Phys. Rev. Lett. **92**, 163002 (2004)

248. E. A. J. Mercatili and R. A. Schmeltzer, Bell System Tech. J. **43**, 1783 (1964)

249. R. A. Bartels, A. Paul, H. Green, H. C. Kapteyn, M. M. Murnane, S. Backus, I. P. Christov, Y. Liu, D. Attwood, and C. Jacobsen: Science **297**, 376 (2002)

250. J. Armstrong, N. A. Bloembergen, J. Ducuing, and P. S. Pershan: Phys. Rev **127**, 1918 (1962)

251. E. A. Gibson, A. Paul, N. Wagner, R. Tobey, S. Backus, I. P. Christov, M. M. Murnane, and H. C. Kapteyn: Phys. Rev. Lett. **92**, 033001 (2004)

252. A. de Bohan, P. Antoine, D. B. Milosević, and B. Piraux: Phys. Rev. Lett. **81**, 1837 (1998)

253. G. Tempea, M. Geissler, and T. Brabec: J. Opt. Soc. Am. B **16**, 669 (1999)

254. A. Baltuska, Th. Udem, M. Ulberacker, M. Hentschel, E. Goulielmakis, Ch. Gohle, R. Holzwarth, V. S. Yakovlev, A. Scrinzi, T. W. Hänsch, and F. Krausz: Nature **421**, 611 (2003)

255. E. Priori, G. Cerullo, M. Nisoli, S. Stagira, S. De Silvestri, P. Villoresi, L. Poletto, P. Ceccherini, C. Altucci, B. Bruzzese and C. de Lisio: Phys. Rev. A **61**, 063801 (2000)

256. G. Sansone, C. Vozzi, S. Stagira, M. Pascolini, L. Poletto, P. Villoresi, G. Tondello, S. Di Silvestri, and M. Nisoli: Phys. Rev. Lett. **92**, 113904 (2004)

257. A. Gürtler, F. Robicheaux, M. J. J. Vrakking, W. J. van der Zande, and L. D. Noordam: Phys. Rev. Lett. **92**, 063901 (2004)

258. K. Lee, Y. H. Cha, M. S. Shin, B. H. Kim, and D. Kim: Phys. Rev. E **67**, 026502 (2003)

259. N. G. Basov, V. Yu. Bychenkov, O. N. Krokhin, M. V. Osipov, A. A. Rupasov, V. P. Silin, G. V. Sklizkov, A. N. Starodub, V. T. Tikhonchuk, and A. S. Shikanov: Sov. Phys. JETP **49**, 1059 (1979)

260. T. J. Englert and E. A. Rinehart: Phys. Rev. A **28**, 1539 (1983)

261. J. Meyer and Y. Zhu: Phys. Fluids **30**, 890 (1987)

262. S.-Y. Chen, A. Maksimchuk, E. Esarey, and D. Umstadter: Phys. Rev. Lett. **84**, 5528 (2000)

263. S. Banerjee, A. R. Valenzuela, R. C. Shah, A. Maksimchuk, and D. Umstadter: Phys. Plasmas **9**, 2393 (2002)

264. K. Ta Phuoc, A. Rousse, M. Pittman, J. P. Rousseau, V. Malka, S. Fritzler, D. Umstadter, and D. Hulin: Phys. Rev. Lett. **91**, 195001 (2003)

265. N. M. Naumova, J. A. Nees, I. V. Sokolov, B. Hou, and G. A. Mourou: Phys. Rev. Lett. **92**, 063902 (2004)

266. A. Rousse, C. Rischel, and J. C. Gauthier: Rev. Mod. Phys. **73**, 17 (2001)

267. This figure has been prepared by Werner A. Hügel, Institut für Angewandte Physik, Universität Karlsruhe (TH), Germany (2003)

268. This figure has been prepared by Oliver D. Mücke, Institut für Angewandte Physik, Universität Karlsruhe (TH), Germany (2003)

269. This figure has been prepared by Thorsten Tritschler, Institut für Angewandte Physik, Universität Karlsruhe (TH), Germany (2004)

270. This figure has been prepared by Klaus Hof, Institut für Angewandte Physik, Universität Karlsruhe (TH), Germany (2004)

Index

Printing: Saladruck, Berlin
Binding: Stein+Lehmann, Berlin